THE ACCESS MANAGER'S HANDBOOK

THE ACCESS MANAGER'S HANDBOOK

A GUIDE FOR MANAGING COMMUNITY TELEVISION

ROBERT S. ORINGEL

SUE MILLER BUSKE

Focal Press
Boston   London

Focal Press is an imprint of Butterworth Publishers.

Library of Congress Cataloging-in-Publication Data

Oringel, Robert S.
  The access manager's handbook.

  Bibliography: p.
  Includes index.
  1. Cable television—Access—Management. I. Buske,
Sue Miller. II. Title.
HE8700.7.075 1986   384.55'47   86-22740
ISBN 0-240-51757-1

Butterworth Publishers
80 Montvale Avenue
Stoneham, MA 02180

10 9 8 7 6 5 4 3 2 1

Printed in the United States of America

This book is dedicated to the fullest implementation of the First Amendment, and the annointing of access and local cable programming as its electronic spokesperson in accomplishing that goal.

CONTENTS

PREFACE

The Access Manager's Handbook was written to fill a need for concise text material in a new and rapidly developing communications field.

All over the United States (and in other countries, such as Canada, Israel, Britain, The Netherlands, and France) cable television access centers are springing up like weeds. Now that access centers are being developed beyond the idea stage, it has become apparent that a new approach to television must be attempted. Because access is a new field of endeavor, and because its creative workforce is mostly volunteer, a new style of manager must be trained and placed in charge of the centers. The success of access cable television depends on this new managerial approach. This book addresses the serious and vital concerns that confront managers in the field.

This is a lean book. It hits hard and pointedly at all of the important issues, and it has not been padded out with extraneous material for its own sake. The primary purpose of this text is to place in the hands of the access manager all of the strategies that have been tried by the authors and others and that have been the focus of several major studies. These studies, although undertaken in the United States, are here reported nationally for perhaps the first time; they have helped make access a success in the communities where it is successful. This book will help access managers avoid reinventing the wheel as they build access centers that work in and for their communities. The appendixes also contain much valuable material for the access manager.

The Access Manager's Handbook was written on an Apple II+ computer, using a word processing program and a spelling-checker program. Nary a pencil, pen, paper, or typewriter was used, except for the paper of the manuscript sent to the publisher, which was printed out using a letter-quality dot-matrix printer. We mention this for two reasons: first, because we later describe the uses of personal computers in the access environment; and second, because personal computers represent a state-of-the-art communications technique.

Concerning the use of personal pronouns in this text, we believe that "he" and "she" are more readable than neutral pronouns, and we have used them randomly.

We sincerely hope that our textbook places or keeps you on the right track in access cable television. We wish you, the reader, and your access center, good luck and much success!

ACKNOWLEDGMENTS

A unique feature of this book is that almost all of the necessary research could be derived from a single source: the archives, literature, and files of the National Federation of Local Cable Programmers (NFLCP). This should surprise no one, since the NFLCP has among its membership most of the talented people who toil in public access, educational access, government access, and local origination cable television. NFLCP has carried out many studies in the field, underwritten by diverse backers. Studies, dissertations, and papers from other sources concerning all phases of access and local origination cable find their way to the NFLCP files at their offices in Washington, DC.

We particularly wish to thank Arlyn Powell, the publisher of Focal Press, for suggesting that we do the book; and we wish to acknowledge the use of *Cable Communication* by Baldwin and McVoy (Prentice-Hall, 1983) as our overall guide to the field. We thank Lane E. Wallace for use of excerpts from her baccalaureate thesis, "Making Public Access Cable Television a Viable Community Communication Resource," (Brown University, 1984). Most of all, we acknowledge the NFLCP for use of the following: articles published in its journal, *Community Television Review;* audio tapes of sessions of its 1984 Denver convention; its "Analysis and Review of Articles of Incorporation and By-Laws," prepared for the Scottsdale (Arizona) Community Cable Corporation; its "A Review of Operating Rules and Procedures," prepared for Montgomery County, Maryland; its *Producer's Primer on Copyright and Fair Use,* prepared for Media General Cable of Fairfax, Virginia by Ernest Sanchez of Arter & Hadden attorneys, Washington, DC; and its "Public Access Administration—Eight Case Studies," prepared for the City of Santa Monica, California, in October 1983. We also thank *Schools and the Cable Revolution* by Jane McKinney, prepared for the Michigan Association of School Boards, 1983. The articles in the July 1985 First Amendment edition of *Community Television Review* form the basis for most of one of our chapters. We thank Professor Michael I. Meyerson, Donald Weightman, Fred Johnson, Professor Robert Perry, Paula Manley, Robert Gurss, Diana Peck, Sidney Dean, and Joe Van Eaton for those articles. Professor Frank R. Jamison and Team 5109, consisting of James D. Zittel, Robert Black, Debra Diederich, Paul Kassab, Bill Jensen, and Sue Wolf—all of Western Michigan University at Kalamazoo—provided us with their Community Access Viewership Study, from which we took material for still another of our chapters. Our final chapter incorporates articles written for *Community Television Review* by Bob Matorin, Paul Steele, Chuck Searcy, Henry Freund, Carol Fites and Frank R. Jamison, Barbara S. Dickinson, and Yolanda Thomas. Appendix A includes material from the survey, "Cable TV Utilization by Colleges and Universities," directed by Robert G. McCartney, Director ITVS, Instructional Television Services, The University of Texas at Arlington. Ha Phan was the researcher on the survey project.

A HISTORY AND OVERVIEW OF ACCESS

THE DEVELOPMENT OF CABLE TELEVISION

Cable television access began only a few scant years ago, and thus it does not yet have the historical background of the Declaration of Independence (and perhaps it never will). Nevertheless, before beginning a discussion of access management, it is important to understand some of television's starting points. This knowledge can then serve as a frame of reference for viewing access. We also wish to begin this historical review in the light of an old cliché: those who ignore history are doomed to repeat it. We stress this because access has not always been a successful venture, and it is a primary aim of this book to eliminate the pitfalls that lead to failure and disillusionment.

In most major US cities, broadcast television was a reality by 1948. But unlike AM radio, which preceded television as the major home entertainment medium, TV is a "line of sight" medium, which means that it was cut off from viewers in areas in which barriers blocked the transmission. There must be a straight shot, with nothing intervening, between the transmitting antenna and the receiving antenna for perfect reception.

Thus, in 1949, the people of Lansford, Pennsylvania—a town about sixty-five miles from Philadelphia—were unable to receive Philadelphia's three TV stations because the Allegheny mountains intervened. An entrepreneur named Robert Tarlton, a local TV sales and service operator, founded Panther Valley Television. This company built a master receiving antenna high atop a mountain peak and delivered what was then called CATV (community antenna TV) to households in the valley below—reputedly for an installation fee of $125 and a $3 monthly charge. Although other similar CATV installations sprang up around the United States at about the same time, the Lansford installation is generally considered to be the first of its kind.

CATV was then extended to communities that were not blocked by natural barriers like mountains, but whose artificial barriers—very tall buildings in large urban centers like New York City—caused "ghosting," or multiple reflections of off-air signals between the source and the receiver. This reflection seriously impaired the quality of the television signal received by these communities.

CATV grew slowly at first, but then it began to flourish. By 1961 there were about 700 CATV systems in the United States and by 1971 there were 2,750 systems.

In the 1970s, the operators of CATV systems started to employ a curious innovation. In order to improve the product being marketed and delivered to subscribers, the CATV operators began to create programming on their own "local origination" channels, and to bring in "imported" signals from distant television stations. Thus, not all cable-delivered channels derived from local off-air television stations; instead the viewer received a choice of programs from many different sources. Occasionally, a few operators even began to deliver uncut movies to their subscribers—on a separate channel, and at an additional cost to the subscriber.

Technological Developments

Aside from the purely historical background, three major technological developments in the same time frame should be mentioned. These advancements spurred the development of cable television to an even higher level.

First there was the introduction of the relatively inexpensive and lightweight "helical scan" video tape recorder (VTR), together with the time base corrector. Used together, these pieces of equipment replaced the extremely bulky, heavy, and vastly more costly "quad" VTR, and made electronic news gathering (ENG) and electronic field production (EFP) possible. This innovation permitted the recorder and video camera to leave the sanctity of the studio and control room and go out into the byways of the community, where the news is made and gathered, and where the community shows its many facets.

The second development was satellite transmission and reception of television programming. This occurrence was brought about by the successful launching of commercial communication satellites such as SAT-COM, COMSTAR, and WESTAR, each with many transponder channels (thirty-five or more). Satellites can transmit programs intercoastally for a fraction of the cost of long-distance coaxial cable or microwave transmission. Thus, cable operators were able to provide entertainment packaging and movie channel networking at a reasonable cost to subscribers.

The third development was 400 megaHertz (mH) cable transmission technology. This expansion in frequency range from 300 mH permitted the cable system, using a dual cable, to be expanded to perhaps 102 channels, rather than the 30 to 35 channels that were previously available. Naturally, with more potential product available, the cable system was more marketable to would-be subscribers.

Types of Satellite-Delivered Programming

Once satellite-delivered programming became available to cable operators, programming networks were developed that have had a profound effect on the cable television industry. At first, some cable operators opposed the introduction of satellite-delivered programming because they considered themselves to be primarily deliverers of off-air television programming. However, when it became evident that the increase in program options constituted a considerable source of new revenue to the operator, all the barriers to the use of satellite programming fell.

Satellite programming can be categorized as follows:

- Pay television. Subscribers can either pay per view—a system that has only been successful for sports events—or pay per month, as typified by HBO (Home Box Office), Showtime, The Movie Channel, Cinemax, Playboy Channel, Home Theater Network, Eros Channel (blue movies), and Bravo, a cultural program service. Cable subscribers pay an additional monthly charge for each of these services that they wish to have, and many subscribers sign up for more than one of them.

  News networks. CNN (Cable News Network) and C-SPAN (Cable-Satellite Public Affairs Network) are examples. These news services are highly rated, and the quality of their news coverage is comparable with that of the broadcast networks.

- Superstations. WTBS in Atlanta, Georgia; WOR in New York City; and WGN in Chicago are examples. These standard broadcast independent stations serve their primary viewing areas twenty-four hours a day, but they can also be received by satellites and all or part of their programming can be fed to subscribers by cable operators. The superstations, which reflect the individual parochial outlooks of their own communities (New York City, Chicago, Atlanta), bring a whole new viewing dimension to subscribers in other communities in the United States.

- Religious networks. CBN (Christian Broadcasting Network), PTL (People That Love, or Praise The Lord) are a couple of examples of the many religious networks on cable. The cable operator is not charged for their use, which perhaps is why there are so many. On the whole, religious networks tend to be fundamentalist and politically conservative. Many emulate "money preachers," such as the Sunday morning radio preachers of old, who exhorted their listeners to "place your left hand on the radio, raise your right hand to God, and don't forget to send your contribution, if you want to be saved."

- Ethnic networks, children's programming networks, and music video, country, and sports networks comprise some of the rest of the programming available to the cable operator—and thus to the cable subscriber—from the satellite transponders.

In the 1980s, cable viewers can expect to receive a wide variety of services: video and audio retransmission—that is, both television and FM radio retransmission of off-air signals; automated video services, with or without an audio background, including character generator transmission of time, weather, news, and delayed stock market reports; locally originated access and other programming that presents local sports events and civic activities; leased channel commercially sponsored material; premium channel pay cable—either per channel by the month, or per view, as in the major boxing events; the special-interest networks, such as PTL; and finally interactive or two-way television. Although still in its infancy, interactive television is destined to become a major force in cable. Using this new technology, the viewer at home can provide input to the system through banking by TV, or ordering groceries from a TV menu, or responding in "real time" to a TV poll about a specific program or subject.

FEDERAL REGULATION OF CABLE DURING THE PAST TWENTY YEARS

In 1966 the Federal Communications Commission (FCC), the US regulatory body in the communications field, asserted its jurisdiction over all cable television by adopting its *Second Report and Order*. But because the cable television industry developed as both a distributor and an originator of programming, FCC regulation was, and still is, confused and confusing. The regulations are currently changing—even as we impatiently wait.

In 1971, under the auspices of the White House Office of Telecommunications Policy, the cable television industry, the broadcasting industry, and representatives of the television program copyright owners—all of whom had been feuding over demands for program copyright usage fees—met and reached a consensus agreement. The agreement outlined the appropriate competitive business relationships that should exist between the two industries and the program copyright owners.

In 1972 the FCC incorporated the consensus agreement into new regulations, and issued the "must carry rules" for (the then) larger cable systems of thirty to thirty-five channels. The order required these systems to provide channel space not only for all of the television stations in their immediate service area, but also for the so-called significantly viewed stations as well.

This order was coupled with a network exclusivity rule, which provided for nonduplication of programs from a distant station, if the same network program was being carried on a local network affiliate.

These were costly requirements that presented several interwoven complexities to many cable operators. Not least among the problems was the fact that when the number of stations available to a community exceeded twelve (the maximum number of VHF channels then available on a tuner with a rotary switch), an interface device called a converter needed to be supplied to each subscriber in order to add additional channels to the subscriber's television receiver.

Also in 1972, the FCC ruled that the larger cable systems must open up access channels. These additional channels were to be free to the subscriber, and were to provide educational access, government access, and public access. Additional leased access channels were required to be available as a common carrier service. These access requirements were imposed upon the cable operators in return for allowing them to use commercial and public broadcasting off-air channels free of charge.

On January 1, 1978, a new copyright law that was passed by the Congress in 1976 went into effect; it significantly updated the original copyright law of 1906. This new law, which now applies not only to print but to electronic media, established specific fees to be paid by the cable and broadcast television industries for the use of signal and of copyrighted programs.

In 1979, a landmark case, referred to as *Midwest Video II*, was brought to court by Midwest Video Corporation, an Arkansas cable company. Midwest successfully fought and challenged the FCC access rules, which contended: (1) that cable operators were required to become common carriers and provide leased access channels open to use by anyone, and (2) that the access rules were beyond the purview of the FCC. The federal courts agreed with Midwest and prohibited the FCC from enforcing its access rules.

But the concept of access had been securely established in the field of cable television by that time. Since 1979, no one seeking a cable television franchise would be foolhardy enough to propose a system to a community without including access channels as a major part of the proposal. It might be noted here, that after proposing grandiose access facilities and receiving the franchise, many cable operators then try to renege and cut back on the promised access channels and access facilities. People in a position to grant cable franchises should make the language of the franchise agreement strong enough to rule out such attempts.

In 1980, the cable industry began to promote a deregulatory federal cable policy that would amend the Communications Act of 1934. Those efforts were successful on October 11, 1984, after four years of hard work on the part of the cable industry and the cities. The Cable Communications Policy Act of 1984 became law on December 29, 1984. The legislation, which was created to provide a uniform regulatory environment for cable communications, will have an uneven impact on access and local programming arrangements throughout the United States. Beyond the fact that the law broadly protects access by its "grandfathering" provisions, very little can be assumed about the new legislation's impact on any local programming or access situation without first considering the particular local cable franchise.

The new legislation is a compromise national policy, and it is of necessity complex. There are certain to be a series of lawsuits initiated by both the regulators and the cable operators that will clarify how the courts view the meaning of the new law.

Since the passage of the new law, the cable industry has issued a number of statements—either through the National Cable Television Association (NCTA) or individual cable operators—proclaiming that the law will lead to a whole range of regulatory changes. These statements, in fact, represent only the industry's hopes and dreams, and no more. The only definitive statement that can be made at this time about the legislation is that no one is very clear about the impact of the new law.

Any changes in existing franchises that are carried out as a response to the new law should be conducted with much deliberation and caution. If the regulatory authorities in your area are considering hasty changes in their local franchise, either at the insistence of the cable operator or anyone else, it should be cause for real concern.

The cable industry has made it clear that they see this new legislation as only the first step in creating the deregulatory environment that they desire. The next step for access and local programming professionals is to educate themselves and their local regulatory officials on the meaning of the law. It should not be very difficult to convince anyone that a period of clarification and education is needed; for example, the law uses the word "service" twenty-five times, with three different meanings at last count.

The remainder of this section discusses the provisions of the legislation that appear at this writing to be most significant to local access and local programming. It cannot be stressed enough, however, that these regulations will be subject to rapid and substantial change.

From an access perspective, the significant feature of the legislation is that, for the first time, Congress has clearly expressed the importance of the public's right to access and has protected it by law. Sections 611c, 624c, and 637a (1) of the law all give local regulatory authorities the ability to enforce and regulate all the provisions that support public, educational, and government access in franchises created before the effective date of the new law. This coverage includes provisions for public, educational, and government access that were not requested in the franchising authority's request for proposals, but that were proposed by the cable operator and included in the franchise.

Section 611a enables the franchising authority to require channels for public, educational, and government access use on both the subscriber and institutional networks of the cable system. This authority includes the right to create rules and procedures governing all three types of access channel.

The provisions for public, government, and educational access clearly represent a very broad endorsement of access by Congress, and should be interpreted as such. Implicit in this endorsement is Congress's recognition of the public's First Amendment rights over those of the cable operators.

Many more aspects of the new federal cable policy law bear discussion: the implications of the obscenity section, contract modification sections, franchise fee sections, rate regulation sections, and many more. All have an impact on access. Ultimately, the force of any particular provision depends on the overall strength of the local regulatory body.

It behooves anyone working in local programming to prepare as well as possible to defend the provisions on local franchises. The National Federation of Local Cable Programmers (NFLCP) has prepared materials that will be useful in understanding the new cable law as it comes into focus in the coming months and years. The NFLCP can also provide information on where to seek legal support if it is necessary.

ACCESS PHILOSOPHY

The philosophy that is arising from use of public access in American communities displays a remarkable parallel to the social philosophies that resulted from the invention of movable type and the printing press by Johann Gutenberg in the 1450s. Before that time, written communication was controlled for the most part by the power brokers—the feudal lords, the wealthy, and the church. They alone could afford the services of scholars or monks to copy and illustrate documents.

FIGURE 1.1 A vintage (1974) public access control room. Courtesy of the NFLCP archives.

Today, the parallel to the domination of broadcast television by large corporations and networks is striking. A few programmers at the broadcast networks decide what all Americans will watch on their televisions—except those viewers who watch access and/or who own video cassette recorders and can acquire rental tapes.

The invention of the printing press heralded a revolution in communications and in global society. In the United States we now take for granted our right to print and read local newspapers; we enjoy the opportunity to read many types of printed information from a broad variety of sources.

In our current era, a similar type of communications revolution is at its beginnings through cable television. Public access television channels can help us meet our neighbors, explore our communities, and address issues of local concern in a totally new fashion. Remember that we are not

FIGURE 1.2   A view from the control room: community producers cablecasting a town hall electronic forum. Courtesy of the NFLCP archives.

talking about television as an entertainment or information medium. In access, the channels are used for a type of community communication. Whereas commercial television is *broad* casting, aimed at a large and diversified audience, access is *narrow* casting, aimed at a small audience with perhaps only one common bond of interest. Whereas broadcasting is one-way and passive, access is two-way and very much participative.

The basic premise of access programming is that the First Amendment grants every American the right to express his or her opinion. For the first time, access provides an electronic forum for both individuals and groups. People can not only express their views about community issues, but can also share valuable community information with their neighbors—indeed, if they want to, they can entertain.

FACTORS FOR SUCCESSFUL ACCESS

What are the factors that make access successful? Here, we will simply identify them; the ones that require an expanded look will be addressed in much greater detail later. Successful access requires:

- A clearly designated access channel or channels. The viewer seeking the access channel should know exactly where on the dial to find it.

- Clearly defined access management, which should be either the cable company, a nonprofit access management corporation, a local library system, a municipality, or some other community-based organization. (This list is in no particular order of preference.)

- An adequate and diversified access funding base. This base may include funds from the cable company, the municipality, or other elements of the community that are willing and able to provide ongoing support for access.

- Adequate and appropriate facilities and equipment, which should include studio and control room facilities, portable location equipment, videotape, videotape editing facilities, and play back facilities.

- Appropriate staff, including a director and as many other people as the funding permits.

- A set of clearly defined operating rules and procedures, in written form, that are available to all access center users.

- A training curriculum that is offered frequently.

- A management plan, an outreach plan, and lots of cooperation among all of the participants, between the access center and its municipality, and between the center and its cable company. Nothing must be left to chance if the success of community access is the goal.

ACCESS AND L/O DEFINED

There are two different types of locally produced programming on cable television systems, and we should clearly define them at this juncture. The two categories are *access* and *local origination,* which is called simply "L/O" and is pronounced like "hello."

L/O is programming produced, directed, and engineered by staff on the payroll of a cable company. Program content is determined by the cable operator's staff and it is most often commercial, includes advertising, and is marketable.

Access is programming produced, directed, and engineered only by community volunteers; in the case of public access, by school system people in educational access and by municipal employees or volunteers in municipal access. The program content is determined by the individual, the group, or the organization that produces the program and it is almost always noncommercial programming that contains no advertising.

Access programming can be subdivided into public access, educational access, and government access. Educational access focuses on education at all levels, from preelementary through graduate school and beyond to adult and continuing education. Course work is taught on and through television. Educational access channels carry school lunch menus, bulletin boards of school events, continuing education courses for adults, and special education for the handicapped. Administration of educational access channels is usually controlled by the local school systems. Educational access operations are often performed by teachers, students, and school administrators assigned to those positions.

Government access channels communicate government information and opinion, and provide a mechanism whereby local government professionals can reach their constituents without having to physically go onto the streets. Police and firemen can update their skills through watching training video programs. Government channels also have bulletin boards listing holidays, trash pickups, and agendas of municipal meetings; they often broadcast live city or county council meetings. Government access operations are performed by local government employees, who may work either full or part time and who may sometimes combine television work with their routine government responsibilities. Cities with well-developed volunteer programs will frequently use volunteers to augment employees.

Lastly, it must be stated that some cable companies are to varying degrees uncomfortable with the concept of community control of what they perceive to be their channel(s). If the distinction between L/O and access has not been made crystal clear in the franchise agreement between the community and the cable company, then the cable company can be expected to try to exert some control over program content on the community channels. Additionally, L/O programming is not always needed, particularly in small communities. It may be far more advantageous to those communities to have well-funded access operations than inadequately funded access and L/O operations. The authors of franchise agreements should carefully consider the requirements of their communities.

ACCESS CENTERS DEFINED

An access center is a place in the community where members can create or learn how to produce programming. A center must have not only the necessary equipment to produce the programs, it must also provide a favorable emotional environment, so that people will be encouraged to transform their concepts into audiovisual displays. An access center combines the functions of a school, a library, and a community center. It resembles a school in that community groups, organizations, and individuals can go to the center to learn to use the tools of television, from a basic to an advanced level. It resembles a library in that people can check out video production equipment (after training courses have been completed) in the same way that books can be checked out of a public library. It resembles a community center in that its environment is friendly, open to all, and is a focal point for community activities.

Most of the access centers in the United States are single facilities—that is, one center serving a community or metropolitan area—but there are multiple public access facilities in such cities as Atlanta, Georgia; Dallas, Texas; Austin, Texas; and Pittsburgh, Pennsylvania.

THE DEVELOPMENT OF LOCAL ORIGINATION

Local origination of programming on cable television systems began at virtually the same time as the CATV systems. L/O grew out of the desire of CATV system operators whose systems had unused channels to create new sources of revenue through local advertising. L/O also represented an attempt on the part of the operators to establish good public relations with their local communities. Finally, L/O was considered a means of creative expression by those who saw the channel(s) as the community's small-town television station.

Who started the first L/O programming is probably in question, but some literature suggests that a Montana operator was doing L/O in the early 1950s, and that a Cumberland, Maryland, operator started in 1953. These early L/O attempts were of course in black and white, used crude methods and untrained staff, did no remote location work, and had no videotape, but their efforts were universally enjoyed in their communities because of their concentration on home-town activities and local children's programming.

By the late 1960s at least thirty cable systems in the United States were doing L/O programming. The limiting factor at that time was the prohibitive cost of the necessary broadcast-level television equipment—the only type of equipment then available. Industrial-level video equipment had not yet come on the scene.

In the early 1970s, as the less expensive industrial-level equipment appeared, L/O began a period of rapid growth. When cable TV entered the urban areas of the United States and the FCC's 1972 rules required L/O in some markets, local origination programming burgeoned. Cable operators in cities like San Diego, San Francisco, and Seattle on the West Coast; Tulsa, Akron, and Toledo in the central United States; and Manhattan and Long Island in New York all made huge investments in L/O equipment and programming. These systems carried sponsored sports programs that were not carried by the broadcast networks.

As might be expected, the sophistication, diversity, and quantity of urban L/O improved more than that of its rural counterparts. Urban L/O included programming about ethnic groups, veterans' programs (particularly about Vietnam vets), and of course an unending variety of children's programming. The appeal of urban L/O also arose from its broadcasts of foreign language programming, city council meetings, debates between local candidates, secondary school activities, and children waving to their parents from the TV screen. In many ways L/O competed with public access.

Since the L/O staffs were producing sponsored, revenue-producing programs, their efforts were a strong influence on the type of programming that cable operators wanted to put on their systems.

Networking of L/O programming was attempted in the early 1970s, despite the fact that a number of incompatible video formats were in use at the time. When, in late 1972, the 3/4-inch videocassette format was introduced and accepted by most of the industry, L/O networking was greatly simplified.

L/O expanded very rapidly from 1970 through 1974. In that year, however, it was decimated—like many other aspects of the cable industry—by the serious economic recession that hit the United States. In addition, the cable industry had overestimated the economic impact that it would have on urban television markets. Many large cable companies tightened their belts and dropped expensive L/O operations entirely unless the companies had been required to maintain it by their franchise agreements. The recession also caused banking interests to cease lending capital for new construction (new builds) to the large multisystem cable operators; bankers regarded such lending as too speculative.

However, smaller systems with no great monetary investment in L/O have continued to flourish, or at least to do well. The advertising potential of L/O is still there, and awaits the marketing skills of those who would use it.

CHAPTER 2

ACCESS MANAGEMENT

ACCESS MANAGEMENT OPTIONS

What are options for access management structure that are open to a community about to embark upon community access?

We know that the citizens of a community must learn to use the tools and the techniques of the television medium before they can begin to produce programming for their own access channel(s), and that these learning experiences often take place where the programs are produced, in a community access center. Access centers in the United States are typically set up and operated by one of four different entities: a cable company, a municipality, a nonprofit access management organization, or another, and perhaps more established, community organization like a church or a public library.

The particular management structure chosen by a community will vary, depending on the provisions of the local cable franchise agreement and on the priority assigned by the franchising authority. The choice also depends on the support offered by community organizations and on whether the access center will be independent of the cable company and/ or the local government.

Since 1972, various access management structures have developed in communities throughout the United States. Describing the experiences of these communities will be useful to communities currently engaged in selecting the most appropriate management structure. We will look first at the advantages and disadvantages inherent in each of the four types of management structure.

Cable Company Access Management

One of the primary advantages of cable company access management is that the access center staff can draw on the experience of other access operations that are managed by the same company. Many cable companies are MSOs (multisystem operators), with established access operations in other cities. The cable company may be more favorably disposed toward access if they are able to exert full control over it.

The primary disadvantage of having an access center operated by a cable company is that its operations are more susceptible to the budgetary whims of the company. Cable companies also tend to allow less community control over program content and facilities than other types of management. Another disadvantage merits serious consideration: cable company policy decisions are often made by company executives from outside the community. During the construction phase of cable company operations when the access center should be built and outfitted, access will often be extremely low on the company's priority list.

Locations where access is managed by a cable company include East Lansing, Michigan; Dubuque, Iowa; Encino, California; Kansas City, Missouri; and Arlington, Massachusetts.

Municipal Access Management

Municipal management systems—that is, access centers operated by a town, city, or county government—usually occur because member(s) of the municipal government want to control the access center or because community access was included in the franchise but neither the cable company nor members of the community wants to take part in its operation.

The advantages of municipal management are the stability that comes with being part of a government body and the steady supply of funding, which can be supported by public taxes. Further, municipal government can usually be relied upon to be responsive to community needs.

The disadvantages arise from political influences. There is danger that access staff will be hired for their political connections instead of their qualifications, and that the center will be used for political propaganda rather than kept at a needed arm's-length from the political machinations of members of local government.

Ann Arbor, Michigan, has a municipally operated access organization.

Nonprofit Access Management Organizations

Nonprofit access organizations are usually incorporated and are normally tax-free. They are identified as "501 (c) (3)" organizations, a reference to that section of the US Revenue code. The advantages include:

- The organization exists solely to assure successful access programming.
- The community tends to have total control over access resources and channels.
- Access operations and programming are more responsive to community needs.
- The cable company does not have control over access channel content.
- The funding will come from a greater number of diverse sources, and thus will not totally depend on any one source.

The disadvantages include:

- The lack of an organizational track record is inherent when an organization starts out.
- The access organization may reflect only a few political viewpoints if a broad cross-section of the community does not become involved.
- The special interests in a community may attempt to take over a weak, new organization to use access for their own ends.

Locations where nonprofit organizations operate access centers include Portland, Oregon; Dayton, Ohio; Tucson, Arizona; Reading, Pennsylvania; Chicago, Illinois; and West Hartford, Connecticut.

Library, School, and Church Access Management

The primary advantage of access management by libraries, schools, or churches is that public access receives the institutional support of these organizations. Also, because of its association with one of these institutions, public access will be perceived as being very stable and secure.

The disadvantage is that access may be dominated by an institutional umbrella that resists innovation, and may thereby attempt to exclude an innovative user whose outlook may be an anathema to a library, a school, or a church. Funding that comes primarily from such an institution may be precarious if access users or prospective users rebel against the suppression of innovation.

Libraries manage access in Bloomington, Indiana; Rome, Georgia; and Fort Wayne, Indiana.

It is important to realize that no one access structure is perfect for every community. Thus, it is desirable to learn as much as possible about the inner workings of at least two centers operating under each type of management structure. This knowledge will help create a more informed discussion when selecting an access management structure.

The reader would be correct to suspect, however, that these authors are biased in favor of the nonprofit corporate access operation.

DEVELOPING THE NONPROFIT CORPORATION

Assuming that a community has opted for an independent access structure, it then needs to set up the structure.

The first order of business for the community would be to incorporate the access organization, primarily to make it an independent legal entity. Incorporation also protects all of the organization's individual corporate members from the legal consequences of excesses or deficiencies attributable to the access company. In most states in the United States, incorporation requires that corporate bylaws and articles of incorporation be written and submitted to the state for approval.

Creating the bylaws and the articles of incorporation that will satisfy everyone usually requires lengthy discussion by the participants in the organization. In practice, these regulations are mostly legalistic jargon, and are seldom looked at once they have been adopted—unless major changes are planned in the structure of the access organization.

Corporate Bylaws

Bylaws typically contain a series of numbered articles that define:

- The name of the corporation.
- The location of its principal office.
- The purpose(s) of the company.
- The membership requirements, in terms of qualifications, classes of members, membership meetings, special meetings, the dues structure, voting procedures, the quorum, and meeting notification.
- The makeup of the governing body—usually termed the board of directors—and their general powers, their number, their tenure, their election, their regular and special meetings, notice of meetings, their quorum, vacancies, age limits, compensation, termination, resignation, the interest of directors, and their liability for debt.

- The officers, the titles of the officers and their duties, the method(s) of officer election and removal, terms of office, compensation, resignation.

- The corporation's standing committees, executive committee, other committees, appointment of committees, terms of office, chairman, vacancies, quorum, resignation, removal.

- Corporate financial transactions in terms of contracts, checks and drafts, deposits, gifts, and grants.

- Corporate accounting systems and reports, such as the accounting system employed, fiscal year, annual report, and the inspection of corporate books.

- Amendments to the bylaws.

- The corporate seal.

Among the more important articles in the bylaws is the description of the organization's purpose. The organization should be described as being organized to operate exclusively for educational, scientific, and charitable purposes, within the meaning of section 501 (c) (3) of the Internal Revenue Code of 1954, or the corresponding provision of any future US Internal Revenue law. Further, the purpose should be to:

- Support local community residents, organizations, and institutions in producing and disseminating noncommercial access programming.

- Monitor and encourage the development of public, educational, and community access programming on the local cable communications system.

- Provide video equipment, resources, facilities, and support services to individuals, organizations, and institutions that desire to produce and disseminate access programming.

- Promote and develop programs for the optimal use of the access channel(s) on the local cable communications system.

- Encourage and develop audiences for the programming on the access channel(s).

- Apply for and receive contributions, grants, donations, and loans of all types from individuals, organizations, public and private corporations, government agencies, and others to support the purposes set forth in the bylaws and articles of incorporation.

There are a number of possibilities for classes of membership, including individuals, organizations, institutions, and corporations. These options are described in Table 2.1.

TABLE 2.1 • OPTIONS FOR BYLAWS

	Options	Suggested Approaches
	Membership options	
Qualifications	Resident	All
	Bona fide interest in access and purposes of corporation	
	Payment of annual dues	
Classes	Individual: regular, students, senior citizens	All options
	Organizational (nonprofit)	
	Institutional	
	Business	
	Patron	

Annual dues	Individual
	    regular $_____ per year
	    student $_____ per year
	    senior citizens $_____ per year
		To be determined

	Organizational $_____ per year	
	Institutional $_____ per year	
	Business $_____ per year	
	Patron $_____ per year	
	Meeting options	
Membership	Annually	All options
	Quarterly	
	Monthly	
Board of directors	Annually	Monthly
	Quarterly	
	Monthly	
Procedures for calling special meetings of membership	By chair of board only	Executive committee or board of directors
	By executive committee	
	By board of directors	
	By petition of majority of members	

(continued)

19

TABLE 2.1 • OPTIONS FOR BYLAWS (continued)

	Options	Suggested Approaches
	Meeting options (continued)	
Procedures for notification meetings	By mail and access channel(s) 15 Days 10 Days 5 Days	15 days By mail and access channel(s)
Procedures for calling special meetings of board of directors	By chair of board only By executive committee	Either option
Procedures for notification of special board of directors meetings	By mail 15 days 7 days	By mail 7 days
Procedures for designation of location of meetings	Determined by chair of board of directors Determined by executive committee	Determined by chair of board of directors
	Voting options	
Definition of membership quorum	25% of total current membership 10% of total current membership 5% of total current membership	Option #1 or #2
Definition of board quorum	⅔ of total board members ¾ of total board members Majority	Majority
Use of proxy votes for membership	Yes No	No
Use of proxy votes for board of directors	Yes No	No
Manner of voting (board)	Voice Ballot Mail	All options
	Committees	
Procedures for establishing special committees	By board of directors Membership	Both options

TABLE 2.1 • **OPTIONS FOR BYLAWS (continued)**

	Options	Suggested Approaches
	Committees (continued)	
Membership of special committees	Any member with interest Only those appointed by board Only those elected by membership	Dependent on committee
	Fiscal authority	
Initiator of contracts on behalf of the corporation	Executive director Chair of the board	Either option
Who may expend funds on behalf of the corporation	Executive director Chair of the board Treasurer of the corporation Any of the above	Executive director Treasurer of corporation
	Bylaws amendments	
Procedure for bylaw amendment	$\frac{2}{3}$ consent of quorum of members $\frac{2}{3}$ consent of quorum of board of directors	$\frac{2}{3}$ consent of quorum of members
When bylaws may be amended	Annual meeting of corporation Special meeting of corporation Regular meeting of corporation Annual meeting of board of directors Regular meeting of board of directors	Annual meeting of the corporation
	Board of directors	
Election or appointment of board of directors	Annual meeting of the membership of the corporation Special meeting of membership of corporation	Annual meeting of corporation
Procedure for filling interim vacancy on board	By individual or body who initially designated board member who vacated By board of directors By membership at special meeting	By individual or body who initially designated board member who vacated By board of directors for other seats

(*continued*)

TABLE 2.1 • OPTIONS FOR BYLAWS (continued)

	Options	Suggested Approaches
	Board of directors (continued)	
Procedure for removing	By directors with ⅔ affirmative vote	Both options
	By membership petition signed by 30% of membership	

Source: Adapted from NFLCP's Recommendations, in a study for Scottsdale, Arizona.

Another article of the bylaws specifies the makeup of the governing body of the organization, usually called its board of directors. When devising the bylaws some of the decisions to be considered are: How many people will be on the board? If there are too few, there may not be a quorum at board meetings when vital corporate business must be discussed and acted upon. If there are too many people, their different opinions may fractionalize the decision-making process. Should the board—at least initially—be elected from among the group of founders, or selected by the city government? Should the board be appointed to represent the various constituencies within the community? (For example, the religious community, the service clubs, the boys and girls clubs.) If the board is elected, what voting method should be adopted? Should the terms of membership on the board be staggered so that members will not all leave office at the same time? Should there be an article that provides for removal and replacement of a board member who misses a specific number of board meetings?

The board should be given specific powers and authority, subject to the provisions of applicable law and the articles of incorporation. These powers should be inclusive of, but not limited to:

- Selection, appointment, and removal of any employees of the corporation; prescription of the duties of the officers; and delegation of such powers to the officers and employees as may be necessary to transact the business of the corporation that is not inconsistent with the bylaws.

- Setting the compensation of employees of the corporation and providing for bonding where it is appropriate.

- Approving expenditures from corporate funds, which will be deposited in a bank or banks designated by the board.

- Approval, at least on an annual basis, of the employment of a competent certified public accountant or auditor to make a detailed examination and audit of the books and accounts of the corporation, and to submit a report in writing to the board.

Other matters for consideration include: What are the titles of the corporate officers? (For example, Secretary, Treasurer, Executive Director.) How are the officers chosen? Should officers be elected by the corporate membership, or should they be elected within the board? What will their duties be? What will their terms of office be? (See Figure 2.1 for an example of cable corporation bylaws.)

Articles of Incorporation

Articles of incorporation for a nonprofit corporation in the state of Maryland (similar conditions are required in other states) must contain:

- An initial statement that the corporation is being incorporated as a nonprofit corporation under the laws of the state.
- The names and addresses of the incorporators (with a minimum number of incorporators specified).
- The proposed name of the corporation.
- The purposes for which the corporation is formed.
- The principal place of business.
- The post office address of the corporation.
- The name and address of the required resident agent of the corporation.
- The provisions of the corporation with regard to the conduct of its affairs: that the corporation is not-for-profit; and that it is not authorized to issue corporate stock, or to declare or distribute dividends, or to undertake any activities not permitted by a corporation exempt under Section 501 (c) (3) of the Internal Revenue Code of 1954 or any state statutes applicable to nonprofit corporations.
- The number of its directors, the names of the initial directors.
- A statement declaring who may be a member of the corporation.
- A procedure for and a designation of power to amend the articles of incorporation.
- The period of duration of the corporation.
- A board of directors indemnification policy statement.
- The procedures for corporate dissolution.
- The corporation's fiscal relationships with state and federal government.

TUCSON COMMUNITY CABLE CORPORATION

BYLAWS

ARTICLE I

Name and Location

SECTION 1. NAME: The name of the corporation, herein called
"corporation," shall be the Tucson Community Cable Corporation.

SECTION 2. LOCATION: The principal office for the location
of business of the corporation shall be within the legal boun-
daries of the City of Tucson, Arizona.

ARTICLE II

Membership

SECTION 1. QUALIFICATIONS: Any person, natural or corporate,
partnership or association interested in promoting the use of
the cable communications system in the Tucson area who is a
resident of the City of Tucson, and who subscribes to or is
a user of the cable system or who makes application for membership
and upon compliance with conditions as may be prescribed by
the Board of Directors, may become an active member of the corporation.

ARTICLE III

Board of Directors

SECTION 1. MEMBERSHIP APPOINTMENT AND RE-APPOINTMENT: The
Board of Directors will consist of fifteen members. Eight members
shall consist of the current eight members of the Tucson Cable
Advisory Commission. The remaining seven members shall be appointed one each by the Mayor and each Councilmember. Upon expira-
tion of these terms, re-appointments for outgoing directors
shall be as provided in Sections 3, 5 and 6 of this Article.

          In addition to the Board of Directors described above,
the City Manager of the City of Tucson or his designee shall
be a non-voting ex-officio member of the Board of Directors.
Similarly, a representative of each entity possessing a cable
communication license for the City of Tucson shall be a non-voting
ex-officio member of the Board of Directors.

FIGURE 2.1   Front page of *Tucson Community Cable Corporation Bylaws.* Courtesy of the
NFLCP files.

ARTICLES OF INCORPORATION

OF

QUOTE...UNQUOTE, Inc.

I, the undersigned, acting as incorporator of a corporation under the New Mexico Nonprofit Corporation Act [53-8-1 to 53-8-99], adopt the following Articles of Incorporation for such corporation:

ARTICLE I

Name

The name of the corporation is QUOTE...UNQUOTE, Inc.

ARTICLE II

Period of Duration

The period of its duration is perpetual.

ARTICLE III

Purposes

The purposes for which the corporation is organized are to operate exclusively for charitable and educational purposes within the meaning of section 501(c)(3) of the Internal Revenue Code of 1954 (or the corresponding provision of any future United States Internal Revenue Law):

1. To develop and promote the concept of public access to existing and future communications media.

2. To establish, maintain and operate one (1) or more media access center(s):

    a. To educate individuals and nonprofit organizations in the use of various media tools and techniques.

    b. To provide individuals and nonprofit organizations with access to various media tools and assistance in their use.

    c. To promote and support the use of various media as vehicles of artistic expression.

    d. To produce programs and other media materials in the public interest.

    e. To establish, maintain and operate a system or systems for the distribution of various media programs and materials, in the public interest.

FIGURE 2.2 Front page of Quote . . . Unquote, Albuquerque, New Mexico, articles of incorporation. Courtesy of the NFLCP files.

The articles of incorporation must be adopted and then signed by the incorporators before a notary. The articles should be reviewed by an attorney licensed in the state of incorporation to assure that they comply with appropriate state provisions.

Samples of bylaws and articles of incorporation are available in a package prepared by the NFLCP, whose address is listed in Appendix B. See Figure 2.2 for one such example. The bylaws and articles written for a particular access corporation must, of course, be tailored to circumstances in the individual communities, but the NFLCP examples should provide some ideas on where to begin the task.

The articles of incorporation together with the bylaws are typically filed by the access corporation with a functionary of the state, who ensures that they are examined for legal form and approved before the company is awarded corporate status. The incorporators, normally the people who start the access organization, may range anywhere from three to a dozen or more people; they usually serve as the first board of directors of the access corporation and continue in that capacity until the first elected board confirms them or replaces them—most often a year after incorporation.

Federal Income Tax Considerations

Considering how difficult it is to acquire access funding, the importance of exemption from federal income taxes cannot be overstated—both for the corporation and for those who wish to support it monetarily.

In order to qualify as an organization that is exempt from federal income taxes under sections 501 (a) and 501 (c) (3) of the Internal Revenue Act of 1954, as amended, the organization must meet all of the following requirements:

- It must be operated exclusively for charitable, educational, scientific, religious, literary, or certain other purposes.
- The net earnings of the organization must not benefit any private shareholder or individual.
- The organization must not propagandize or otherwise attempt to influence legislation as a substantial part of its activities.
- The organization must not participate in or intervene in any political campaign on behalf of any candidate for public office.

In addition to these regulations, which are stated in section 501 (c) (3) of the code, the IRS has added the following requirements:

- The organization must not be organized or operated for the benefit of private interests.

- The organization must not engage in racially discriminatory activities.

How do these stipulations relate to the articles and bylaws of an access corporation? An exempt organization must meet both an operational test and an organizational test. It will satisfy the operational test if its activities conform to the described requirements. It will satisfy the organizational test if its articles of incorporation meet all of the following requirements:

- The articles must limit the organization's purposes to one or more of the exempt purposes described in section 501 (c) (3) of the code. The articles may not empower the organization to engage, other than as an insubstantial part of its operations, in activities that do not further one or more exempt purposes. If the articles empower the organization to engage in such activities, then the organization will not meet the organizational test, even if the articles state elsewhere that it is organized solely for exempt purposes.

- The articles must state that the net earnings of the organization will not benefit the members, trustees, directors, officers, or other private persons.

- The articles must permanently dedicate the assets of the organization to an exempt purpose. The articles must therefore provide that, upon dissolution of the organization, its assets will be distributed for an exempt purpose or will be given to the federal, state, or local government for a public purpose.

- The articles may not expressly empower the organization to devote more than an insubstantial part of its activities to attempting to influence legislation by propaganda or other means, or to directly or indirectly participate or intervene in any political campaign on behalf of, or in opposition to, any candidate for public office.

Although an organization's bylaws do not have to meet any specific requirements for the purpose of establishing an exempt status, be aware that any statements in the bylaws about the activities and purposes of the organization should be consistent with its articles of incorporation. When applying for exempt status from the IRS, the access organization must submit copies of both its bylaws and its articles of incorporation.

DEVELOPING RULES AND PROCEDURES

At the outset of this discussion, the reader may question the need for rules and procedures in a communications medium that purports to be a model of free expression and free exchange of ideas. Let us therefore state, unequivocally, that rules are indeed necessary. The truth of the matter is that if there is to be freedom, there must be order. The absence of order is chaos, and chaos spells the end of freedom. For this reason, we spell out rules that ensure that public access television will operate in a fair, systematic, and equitable fashion. Further, if the access center is to avoid domination by a particular individual or group, deterioration of equipment, disarray or theft of facilities—and all the while guarantee access to access for all—there must be rules.

Having established the rationale for an access center's operating rules and procedures, let us spell out a context for them.

- The rules should be written clearly and understandably, and should be available to all users, probably in the form of a manual.

- The rules for a specific center may best be devised by a group that includes the director of the access center, the members of the access corporation board, and perhaps representative(s) of the franchising authority and the cable company.

- The rules should spell out in detail the responsibilities of the people who will be using the facilities. The facilities typically include portable location equipment (porta-paks) and in some cases mobile control room vans, the studio and control room, the editing suite(s), and the access channel itself. Training requirements, equipment reservation, checking in and checking out procedures, and restrictions on use of tapes and equipment must be clearly described.

- The rules should delineate a system of penalties—whether monetary or in the form of restrictions on the use of facilities—imposed for infractions of the rules. The access center must operate much like a library, where the next user depends on the responsible actions of the previous user.

- The rules and procedures document must be kept a *living* document. Changes, additions, and deletions should be possible from time to time as the access center and its users mature. Many of these changes will not be apparent when the document is originally drafted, so an amendment process must be included.

- Guidelines for program production should be included, along with procedures for submitting a program for cablecast, and examples of all of the forms to be completed by users to request access center services.

- The rules should describe procedures for enrolling in training courses, outlines of those courses, and locations where they are taught.

- The rights of users concerning program content should be clearly spelled out.

- Fees, if any, that are charged for use of the studio or editing facilities, training, or the use of portable equipment or tape should be clearly stated. Although fees may promote more careful handling of equipment, they clearly will discourage overall equipment use. Fees will limit the number of people who can take advantage of access program production, particularly among youth, senior citizens, and low-income groups. These groups, ironically, should be among the largest users of access—if only because they have the time to do it. We recommend a no-fee, no-deposit system.

THREE IMPORTANT ISSUES

Judging from the previous experiences of many communities, at least three issues are guaranteed to confront the group that convenes to construct the rules and procedures. The three issues are obscenity, prime time scheduling, and program quality.

Obscenity

Obscenity, more often than not, is really a nonissue in public access television. The issue, when examined closely, turns out to be a question of compliance with community standards more than anything else.

Most access programming is pretaped, so censorship could be easily exercised by the access center staff, the cable company, or by the center board of directors. But please be strongly advised, whatever your personal inclinations are, that access users are protected against "prior restraint" of their material by the First Amendment to the US Constitution and by provisions in the Cable Communications Policy Act of 1984.

The solution to the apparent dilemma, therefore, is not in censorship, but in airing the purportedly offensive program at a time when it will offend the least number of people—say, after 10:00 PM, or perhaps even 3:00 AM. Realize, too, that no program will offend everyone, and that some will leap to the defense of what others call obscene.

It is important that your access group does not waste the many hours that some community access groups have wasted by agonizing over rules to control a situation that may never occur, or if it does, can be managed by judicious scheduling.

Prime Time

Almost every producer wants his or her program aired when the maximum number of viewers are available and watching. That period is generally between 8:00 PM and 11:00 PM, the hours normally referred to as prime time. Rules that guarantee balance and equitability—sometimes by using a rotating program schedule—often solve this problem to the extent that it can be solved. Access centers frequently give priority in prime time to series programs. This is done on the theory that access producers who provide the community with ongoing programming deserve the best showcase for their programs. The rules and procedures should contain methods for scheduling series programs, and for dealing with first time and ongoing series productions.

Quality

For a long time, the broadcasting industry has had policies that define air-worthiness in terms of the technical "broadcast quality" of programs, both in radio and television. In the interest of maintaining a quality signal, broadcast stations will not air programs that are below specific technical standards, with the notable exception of news film or video inserts that are irreplaceable—such as the 8 millimeter amateur film of the assassination of President John F. Kennedy. The FCC, too, maintains technical standards to which broadcast stations must adhere.

In public access television, we are—or we certainly should be—less constrictive about technical quality. We should always aim as high as possible for technical quality, but we must also recognize that our people are not professionals and that our equipment is at best industrial grade, and at worst consumer grade. It is therefore not always possible to have the best picture or sound signal. Also, in access the message should be considered at least as important as the medium, except in a situation where the message is totally distorted by a very flawed medium. We should never reject a program simply because it fails to come up to arbitrary technical standards, even if we must append a disclaimer to the program stating that it was the best that could be done under specific stated circumstances. All the while, we should be trying to get better and better in a technical sense.

THE RULES AND PROCEDURES MANUAL

The procedures manual should begin with a description of the access corporation, including its founders, its current board of directors, and its officers. It is also a good idea to present a definitive statement that describes the access philosophy. The manual should have a clear table of contents that organizes the following text. Location of the facilities, hours of operation, a description of the available facilities, and eligibility to use them should be included as well. Figure 3.1 shows the cover of Atlanta's access manual.

Some of the necessary types of rules and procedures are examined in detail below.

Procedures for Studio Use

The section of the rules and procedures manual that discusses studio use should define who can use the studio and control room, how studio production time is scheduled, and specifically who does the scheduling. In most instances, any individual who has received studio use training in the access center and who has passed a hands-on certification test may request scheduled time to do a studio production of an approved program. Approval does not in any way imply censorship; rather, it describes a program idea that has been discussed with the center director or her designate to insure full consideration of all of the production factors that will point the program in the direction of success.

Time is usually scheduled on a first come, first served basis, with some accommodations made and time slots reserved for series programs. Many access centers use a form that must be completed and submitted at least forty-eight hours in advance of intended studio use (see Figure 3.2). Usually the program producer is responsible for recruiting an operating crew, but in many cases the staff of the access center will maintain a file and serve as a central clearing house for potential crew members. All crew

ACCESSING ACCESS:
*Becoming part
of
Community Television*

*A handbook of
Cable Atlanta, Inc.'s
Center for
Community Television*

*1018 West Peachtree Street, NW
Atlanta, Georgia 30309
(404) 874-8000*

FIGURE 3.1 *Accessing Access,* cover of the Atlanta, Georgia, Center for Community TV access manual. Courtesy of the NFLCP files.

members must be certified in studio production, and should be expected to help in both studio setup before a production and in striking the set and cleaning up the studio after a production.

Procedures for Portable Equipment Use

An introduction to the portable video camera and the portable video cassette recorder is often the first training experience that is given to the beginner in public access television. As with training in the studio/control

```
___STUDIO                          REQUESTED DAY:_____
                                          DATE:_____/_____/_____
___EDITING                         TIME-FROM:_____TO:_____
                                          --REMOTE---
___REMOTE                          RETURN DATE:____/_____/_____
                                   TIME OF:  Pick-up:_____
                                             Return:_____
                   --FACILITIES REQUEST--

NAME:_____  USER NUMBER:_____
     (print)
ORGANIZATION:_____

HOME PHONE:_____  WORK PHONE:_____

PROGRAM TITLE:_____

TAPE REQUIRED (Length):_____  TAPE #'S:_____

                     --FACILITIES--
_____

  STUDIO                            EDITING

  ___Character Generator            Source Machine:
  ___Audio Tape Recorder            ___3/4"
  ___Camera(s)                      ___1/2"
  ___Microphone(s)                  ___Character Generator
                                    ___Audio Tape Recorder
_____ REMOTE _____

 --1/2"--              --3/4"--

  ___Camera             ___Camera            ___Switcher
  ___VTR                ___VTR               ___Wireless HS
  ___Battery(s)         ___Battery(s)        ___Video Cables
  ___Power Supply       ___Battery Belt      ___(for switcher)
  ___Tripod             ___Power Supply      ___Cube Tap(s)
  ___Monitor            ___Tripod            ___3-2 Prong AC Adapt.
  ___Monitor Cables     ___Monitor           ___AC Ext. Cord(s)
  ___Lights             ___Monitor Cables
  ___Microphone & Adapt.___Lights            NOTE:  Indicate # of
  ___Headphones         ___Microphone(s)     items required when
  ___Cube Tap           ___Mike Cable(s)     appropriate.
  ___3-2 Prong Adapt.   ___Headphones
  ___AC Ext. Cord(s)
  ___Mike Cable
_____

  I agree to be financially responsible for all equipment in my possession

  SIGNED:_____  Date of Request____/_____/_____
----------------------------FOR OFFICE USE ONLY----------------------
DATE APPROVED:____/_____/_____     By:_____

Remote Equipment- Checked out by:_____  Checked in by:_____
```

FIGURE 3.2 Studio use application form.

room and editing equipment, instruction is followed with a certification test. Again, a special completed form is required before training commences, and another form must be completed and submitted in advance before the portable gear can be checked out. Requests are usually honored on a first come, first served basis. The checkout form for borrowing portable equipment is a simple but binding contract between the access center or the cable company and the user; it concerns the user's monetary responsibility for the equipment (see Figure 3.3). This responsibility must be made very clear to the user.

On the request form, the user, by his signature, accepts complete responsibility for damage to the equipment, except for normal wear and tear. Often the access center will carry an insurance policy that covers the equipment when it is being used by staff and volunteers. This insurance is described in Chapter 6.

Many access centers require users to arrive punctually at the specified time of portable equipment checkout, because the valuable time of staff members or technicians is involved. Some access centers record an identification number for each item of equipment on the checkout contract. All centers confirm the equipment's working order, which makes its condition a joint responsibility shared between the staff member and the user, and then indicate the day and hour of its required return.

The user signs for the equipment and also signs a statement of compliance. This statement confirms that the user is familiar with the access center rules and with the access organization's parameters on program content to be taped. The user agrees that no advertising or commercial material, no obscene material, and no lottery information will be videotaped.

The producer/user agrees to accept responsibility for any disputes that may arise by agreeing "to indemnify and hold harmless the access center, the cable company and the community, from any and all liability, or other injury including all reasonable costs of defending claims and litigation arising from or in connection with, claims for failure to comply with any applicable laws, rules, regulations or other requirements of local, state or federal authorities; for claims of libel, slander, invasion of privacy, or infringement of common law or statuatory copyright; for unauthorized use of trademark, trade name, or service mark; and for any other injury or damage in law or equity which claims result from the user/producer's use of the access channel(s)."

When the equipment is returned, the technician on duty and the user examine and test it to insure that it is all there and is functioning properly. A user who returns equipment late may be given a written warning or, alternatively, may be fined, either monetarily or by temporarily suspending his further use of the access facilities.

DATE_____

NAME_____ PHONE_____

ADDRESS_____ DR. LIC.#_____

FROM_____ TO _____
     DATE   TIME   DATE   TIME

PORTAPAK I

Portable Camera 585
Viewfinder   586
AC Adaptor   596
Record Deck   579 _____

PORTAPAK II

Portable Camera 581
Viewfinder   580
AC Adaptor   577
Record Deck   598 _____

PORTAPAK III

Portable Camera 591
Viewfinder   590
AC Adaptor   583
Record Deck   585 _____

PORTAPAK IV

Portable Camera 595
Viewfinder   594
AC Adaptor   577
Record Deck   589 _____

Battery	578_____	584_____	588_____	597_____	Tape#_____
Mini Pro	516_____	517_____	518_____	609_____	
	610_____	611_____			_____
Tripod	558_____	559_____	612_____	613_____	Other_____
	614_____	615_____			
LT. Stand	519_____	520_____	522_____	560_____	_____
	561_____	603_____	604_____	605_____	
	606_____	607_____	608_____		

Positive I.D. is required. Minors must be accompained by a responsible adult. I agree to return this equipment to the Allen County Public Library at the Telecommunication office. I also agree to pay any and all costs incurred for damage or loss from negligence or abuse while the equipment is signed out to me.

Signature_____

There is a $1.00 per/hour fine for overdue equipment. Due time is designated on this form. User will not be charged for those hours the library is closed.

Checked Out By_____Date_____

Checked In By _____Date_____

Comments_____

FIGURE 3.3 Portable equipment checkout form.

Procedures for Mobile Control Room Van Use

The mobile van is a multicamera control room on wheels, which often contains more camera facilities than the access studio. Its use is normally limited to large location productions, such as live music programs and sports events. One or more of the access center staff members operate the van, in addition to the usual complement of access program crew volunteers. In cities with more than one access center, there is often only one van serving all of the centers.

Use of the van is therefore carefully scheduled. Often the van can only be checked out on the recommendation of the access center manager, who, with the program producer, determines that the van's facilities are needed to make a particular program. Advance scheduling is obviously a necessity. The studio/control room request form is often used for the van as well.

Procedures for Editing Suite Use

Training in videotape editing is normally available to program producers who have completed a portable equipment or studio production class. At some access facilities training is restricted to program producers with a "show tape in hand," simply because time spent on the usually limited editing facilities is at a premium. Certification in editing is gained through this training, and is required for use of the equipment. Frequently, users must submit a completed request for editing time form at least forty-eight hours in advance of the intended use, but since the facility is normally in great demand, most access centers suggest that the request be made at least seven to ten days in advance to insure the availability of the requested editing time slot. Time during the day is usually easier to get than during the evening, since most program producers either have a job or are in school in the daytime. Some access centers set a monthly time limit on the number of hours that editing equipment may be used by an individual, with a waiver available for special cases.

Procedures for Access Channel Use

The individuals, organizations, or groups who wish to schedule time on an access channel, either for a tape that was produced using the access center facilities or for a tape produced elsewhere, must complete a cablecast request form (see Figure 3.4). This form specifies the name and subject of the program, the name of the individual or group submitting the tape for cablecast, the name of the program producer, the length of the program, and whether it contains an obligatory thirty seconds of introductory tape at the front end. This thirty-second introduction includes color bars, audio description and tone, slate, and countdown. The form also asks for

AUSTIN ACCESS PROGRAM CONTRACT

PROGRAM TITLE: _____

EQUIPMENT USED: ( ) ACCESS   ( ) OTHER   ( ) COMBINATION
                   ( ) STUDIO   ( ) FIELD   ( ) COMBINATION

PROGRAM TYPE: ( ) SERIES   ( ) NON-SERIES
                   ( ) LOCALLY PRODUCED   ( ) NON-LOCALLY-PRODUCED
                   ( ) ARTS/ENTERTAINMENT   ( ) RELIGIOUS   ( ) SPORTS
                   ( ) PUBLIC AFFAIRS   ( ) EDUCATIONAL   ( ) OTHER

PLEASE PRINT:

Name:_____ Phone_____

Address:_____ City_____ State_____

Sponsor:_____   Zip_____

Channel Space Needed: ( ) 1/2 hour ( ) hour ( ) Other_____

Would you like your program repeated? ( ) yes ( ) no   Comments:_____

_____

Does the program contain copyrighted material?   ( ) yes   ( ) no

Is use of the copyrighted material a "fair use" under copyright law?   ( ) yes   ( ) no

Have you submitted a release for copyrighted material in the program with this application   ( ) yes   ( ) no

Permission to copy this program for limited purposes _____ (initial).

Please list crew members: _____

_____

Please provide one sentence description of program: _____

_____

SPECIAL INSTRUCTIONS:

Program must be delivered to the Programming Department no later than noon, one week ahead of its scheduled cablecast unless special arrangements are made.

Programs must meet minimum technical requirements (list available at Austin Access) and be properly labeled (labels available at Austin Access).

I, the undersigned warrant and represent to Austin Access and Austin Community Television (ACTV) that the above program submitted by me contains none of the following:

1. any material which violates state or federal law relating to obsenity;

2. any material which is libelous, slanderous or an unlawful invasion of privacy;

3. any advertising or material which promotes any commercial product or service;

4. any use of material which violates copyright law;

5. any material contrary to local, state or federal laws, regulations, procedures and policy;

6. any material which appeals for funds.

These warranties and representations are made by me in order that this program be cablecast free of charge on Austin's public access channels. I agree, further, to indemnify and save harmless ACTV, the City of Austin, Austin CableVision and any of their employees, officers, Boards of Directors, stockholders, etc., from any and all claims, demands, damages or other liabilities which may be made against or arise out of the cablecasting of the program submitted by me whether or not the program has been reviewed by ACTV prior to cablecast. I further agree to pay ACTV, the City of Austin and/or Austin CableVision all legal fees and expenses incurred by this program in connection with any legal proceedings concerning its cablecast, as such legal fees and expenses arise. I am aware that Section 639 of the Federal Cable Communications Policy Act of 1984 provides that

"Whoever transmits over any cable system any matter which is obscene or otherwise unprotected by the Constitution of the United States shall be fined not more than $10,000 or imprisoned not more than 2 years, or both."

I also understand that this program application is subject to the Open Records Act (Article 6252-17a, V.T.C.S.), and may be released as a public document.

Signature_____ Date_____ 19_____

Additional Comments: _____

_____

_____

TO BE COMPLETED BY ACCESS PROGRAMMING STAFF:

CUETIME ____:____   PROGRAM # _____   STAFF INITIAL _____

RUNTIME ____:____   PRODUCER # _____   DATE _____

1/85

FIGURE 3.4 Access channel time request form.

the requested date and time of day of the initial cablecast. If the program is produced at the access center, most centers require that it be completed before scheduling it for cablecast; usually, the completed tape must be submitted from two to seven days preceding the scheduled cablecast.

THE PROGRAM PRODUCER'S RIGHTS

Within the access world there has been some confusion over who owns the rights to access programs. It is commonly thought that the program producer holds the rights, and recent cable legislation seems to confirm this belief.

Note that what we discuss here concerns the *content* of the program, not the tape that holds the program. If ownership of rights is not clearly defined in the operating rules manual and is not rigidly adhered to by the access center and the cable company, then legal actions that are—at best— difficult to defend against can be expected from disgruntled access producers.

Simply stated, *the producers of public access programs retain the rights to the contents of their programs.* By submitting the program for cablecasting, or by using access center facilities for production of the program, the public access user does not relinquish any rights to the program content.

The signature of the access user on a public access cablecast request form only gives the access center and the cable company limited rights to cablecast the program one or more times. If the center or the cable company has supplied the videotape, then the producer is entitled to a copy of the tape if he or she pays for it.

Under no circumstances, should the access center or the cable company:

- Duplicate a public access producer's tape, or any portion thereof, without prior permission from the producer, except when a public access sampler tape is being compiled to help promote the concept of public access. An access producer who does not want a portion of his or her tape used in this fashion must state so in writing when the tape is submitted for cablecast.

- Use a public access tape on any commercial or nonaccess channel, or distribute the tape commercially.

- Make a copy for a third party without the express permission of the producer, except when required by a regulatory body or a court of appropriate jurisdiction. In any event, the public access producer must be notified in writing before any such action is taken.

Another related area concerns the access user's lack of right to commercially distribute his or her tape. Most access centers operate under the belief that since their equipment, training, and facilities are provided free of charge, the access producer should not make personal monetary gains from commercial distribution of the tape. If the producer does earn money from the program, most access centers ask to be compensated for the use of the equipment at a rate equal to the lowest commercial rate available in that city for rental of like facilities.

USE OF VIDEOTAPE STOCK

Access centers normally maintain a supply of videotapes for use by access program producers. In some centers, these work tapes are only loaned to center members, while nonmembers are encouraged to purchase their own tapes. In either case, work tapes in use for a production are kept at the center until the production is completed and a master tape is prepared. Work tapes are then normally recycled into stock—that is, erased (degaussed) and shelved, sometimes after a specific time period, say thirty days. This method insures that an adequate supply of tape is available at all times, without vast expenditures for tape stock. The negative side of this procedure is that the user occasionally and inadvertently gets an erased tape that has areas that have been "stretched" by repeated shuttling back and forth during the editing process.

Producers who wish to reserve work tapes for more than the allotted thirty days must request approval from the access manager.

TRAINING PROCEDURES

Training is a major function of every access center, and is normally divided into the three categories necessary to get video production under way: basic, studio/control room, and videotape editing or postproduction. At many centers, these categories are further subdivided into basic and advanced levels of each category. Additional seminars often include: the use of zoom lenses, studio lighting, stagecraft (set-making), and the mobile production van. Special workshops are held to discuss script writing, advanced audio, stage makeup, and other pertinent subjects.

The procedures inherent in this training include application forms, course descriptions, places and times for registration, places and times for the courses, penalties (if any) for nonappearance or partial appearance at courses, and end-of-course testing and certification.

GUIDELINES FOR PROGRAM PRODUCTION _____

One of the functions of the access manager and her staff is to gently lead the neophyte through the maze of access program production. This is done through one or more conferences with the producer, in which the proposed production is discussed based on a completed Program Proposal form. The conferences wed the producer's ideas to the center staff member's skills and knowledge. Production values are discussed, as well as the technical standards adhered to by the center. If the production ideas generate a program series, the center's procedures for series programs are also discussed.

APPEALS PROCEDURES _____

Because it deals with the sensibilities of diverse people, the access center will occasionally be involved in disputes or disagreements between access producers and the center, or between access producers and the cable company. Most disputes involve time scheduled or not scheduled for production on the access channel. Therefore, an appeals procedure should be set up in advance, so that disagreements do not deteriorate into fistfights or litigation.

Since there is a wide range in access center setups, no one system of appeals would fit all, but some general statements regarding appeals can be made.

A two- or three-level appeal system appears adequate. The first level would be an appeal by the producer or another user to the access center manager or to the operations manager of the cable company, whichever is pertinent. If the complainant receives no satisfaction, the second level of appeal would be to either the board of the access corporation or the general manager of the cable company. The final appeal would be to the franchising authority.

The appeals method should require a written statement of complaint; a conference within a short, stated time after receipt of the complaint; and a written response to the complainant within a short, stated time, for each level of complaint. Remember that most people feel that justice delayed is justice denied, so rapid conclusion to a complainant's appeals is necessary.

THE ACCESS CENTER STAFF

This chapter discusses the day-to-day work that is required to operate a successful access center. We describe the different jobs according to the needs of the center instead of by the job title. We take this approach, rather than one that outlines job titles and the concurrent duties of those titles, because we cannot know how many staff members your center can afford to have on its payroll, or how many volunteers or interns will be available. We do know, however, what jobs need to be done on a continuing basis to get the center started and to keep it moving with a high level of success.

QUALIFICATIONS OF THE ACCESS MANAGER

There must be, first of all, a full-time access center manager or boss. The job title will vary from center to center; although often labeled executive director or access coordinator, the title may range to the other end of the spectrum, where the manager is just called "hey, you."

However, the important thing is that the access manager has most of the following attributes. The person must be a firm but not unyielding manager. The people using the center are volunteers, not paid employees, so the manager must be able to cajole and command without seeming to demand. He should be able to imbue the rest of the staff with confidence that he is a strong and able facilitator. The other staff members, both paid and volunteer, must be made to understand by the way that he interacts with them that he has their personal interests and concerns uppermost, and that he is not just going through the motions. They must know that he will work tirelessly to see that they get whatever is needed in the way of encouragement, funding, or equipment, so that they can do their jobs and feel good about those jobs.

It is important, too, for the manager to understand that he is to get his "psychic salary" by inspiring the same sense of satisfaction in the other members of the staff, rather than by seeing his name credits on programs. Above all, he should know that he is *not* the executive producer of the center's programs, but merely a facilitator who willingly backs away when his expertise is dispensed and accepted—or, indeed, rejected—and lets others take production credits.

The specific managerial requirements involve the day-to-day operation of the center. They include dealing with the center's board of directors—often a foreboding task, considering that the manager must live with a company policy made by committee. The requirements also encompass day-to-day dealings with the cable company's operations manger or director of community programming, and periodic dealings with the managerial staff of the municipality.

When examining the duties and responsibilities of various access co-ordinators, general managers, and executive directors, we discovered a very complete job description that was compiled by the Portland Cable Access Corporation in Portland, Oregon. This corporation is a nonprofit access management organization, and thus the description is written from a nonprofit perspective. However, we believe that many, if not all, of the duties and responsibilities described in the following list are clearly applicable to an individual managing access for a cable company, a community organization, and a municipality, as well as for a nonprofit corporation.

Some of the tasks typically performed by the access manager include:

- Define manpower needs and specifications; recruit, hire, train, supervise, evaluate, reward, discipline, and perform other necessary personnel management functions with subordinate staff, both paid and volunteer.
- Carry out approved policies, procedures, goals, and objectives for the corporation; make recommendations to the board for additional statements of policy direction and/or additional or revised goals as appropriate.
- Establish, furnish, and maintain necessary business operating facilities, including office and delivery equipment and supplies.
- Establish and maintain an outreach campaign to community groups, businesses, nonprofit organizations, and all potential users of cable access in an effort to recruit programmers, viewers, and volunteers.

- Monitor telecommunications legislation, especially as it affects cable access programming; make recommendations to the board for appropriate action.

- Establish and maintain the support systems necessary to effectively provide adequate background, training, and a conducive environment to individuals or groups wishing to do programming on the cable access channels, or with the cable access facilities under the jurisdiction of the access corporation.

- Develop alternative and additional sources of revenue, such as federal, state, and/or private philanthropic grants.

- Serve as chief purchasing and business agent for the access corporation. Prepare and submit financial and activity reports to the board of directors, the auditors, and the public.

- Administer budget and work plans approved by the board. Provide the board with background information, technical and administrative knowledge, and alternative recommendations necessary for their decision-making process.

- Attend and conduct meetings.

- Perform related duties and assume additional responsibilities as needed by the growing access corporation.

The following list of knowledge, skills, abilities, and other desirable characteristics clearly indicates the type of individual who will best fulfill the role of access manager. This list is also excerpted from a compilation of the Portland Community Access Corporation.

- Knowledge and understanding of the background, present use, and potential of cable access.

- Knowledge of the equipment, systems, and facilities involved in cable access programming and cablecasting technology.

- Knowledge of and ability to work within public and political processes.

- Knowledge of personnel management policies and practices, including legal guidelines and restrictions.

- Knowledge of volunteers: recruitment, training, and motivation.

- Skill in public speaking, teaching, training, and persuasion techniques.

- Skill in written communications and report writing.

- Skill in fundraising, budget development, and management of fiscal resources and financial reporting systems.

- Skill in management and leadership techniques.

- Ability to deal diplomatically and tactfully with diverse public and political representatives.

- Ability to function calmly and effectively during periods of high pressure and stress.

- Ability to maintain a high level of energy and enthusiasm throughout the normal work day, evening meetings, and long work weeks.

- Ability to recruit, train, supervise, and motivate both paid and unpaid staff.

- Ability to analyze complex problems or situations, to conduct or direct appropriate research, and to recommend effective courses of action. Willingness to render judgment and make decisions within the scope of stated responsibilities.

- Strong sense of social conscience: a desire to positively affect the community and improve community communications through cable access.

- Ability to conduct oneself, both on and off the job, in a manner befitting a top-level executive.

In a worst-case situation, all of the other skills and duties necessary to the operation of the center may be done by the manager—such as the communication skills necessary to perform television operations, and the training skills needed to teach television operations, and the fundraising and outreach duties that provide the money and the center's volunteers. Or, these tasks may be delegated to other people, including volunteers, which would be the best-case situation. The size of the available staff determines who does what. Although the duties mentioned, such as training, may be delegated to other people, the access center manager has the ultimate responsibility for them.

The manager's job is complex, and requires a singular person with intense devotion to public access television; however, it is usually a much lower-paid job than a counterpart position in commercial broadcasting.

There is a constant controversy in access circles over whether or not someone with a broadcasting background is an asset or a liability to a public access operation. Many people argue that the experienced broadcaster may have become tainted by slick commercial television. In fact, this concern—like that over the obscenity issue—is a tempest in a teapot. Some former broadcasters will be terrible for access, and some will be superb; the determining factor will be the worth of the individual and the dedication to access shown by that individual. The important issue is not where that person was last employed, or even whether access is being used as a stepping stone into, or back into, broadcasting.

OTHER ACCESS CENTER JOBS

Aside from the managerial position, there are other necessary jobs at the center.

A preproduction planning specialist is needed to confer with volunteer program producers and add production expertise to their program ideas (see Figures 4.1, 4.2, and 4.3). This person should have a thorough knowledge of the use of the television medium. Senior university interns working on degrees in communication are valuable in this position. These interns are recruited by the access manager through cooperation with the schools of communication of the local colleges and universities. The interns devote a specified number of hours to the center in a dual role. They help the volunteers as suggested above, and at the same time they work on television production projects approved by their schools to earn college credit.

It is also vital for the access center to have someone to do scheduling. Training courses, with their times and places, must be scheduled and the students assigned, including follow-up after the course is completed. Location and studio shoots—including the times and places for loaning out

FIGURE 4.1 Access facilitator conferring with volunteer producers on preproduction planning. Courtesy of the NFLCP photo files.

FIGURE 4.2 Access facilitator assists community producer with camera operations. Courtesy of the NFLCP photo files.

FIGURE 4.3 Access facilitator describes use of control room equipment to a group of trainees. Courtesy of the NFLCP.

and checking in portable equipment—and production time in the studio must all be scheduled. And, of course, time on the access channel(s) must be coordinated and scheduled accurately.

Someone is needed to do secretarial and clerical duties: typing letters, photocopying forms and scripts, and keeping the access office operationally neat.

A coordinator of volunteers, a production coordinator, an intern co-ordinator, and a budget and finance planner are required at the center. Budget and finance planning is essential to a well-run center, and in small communities these tasks can often by done by the access organization's treasurer in conjunction with his disbursing duties, or by a committee headed by the treasurer. How many paid positions there are at the center depends totally upon the operating budget. Salaries typically make up a large percentage of a center's operating budget.

The professional technician's duties are sometimes performed by people on the payroll of the cable operator, and sometimes by staff on the payroll of the access management. The center needs a technician to play back tapes from a VCR to the modulator that feeds the programs to the access channel. This person must be on duty whenever there are access programs on the channel(s). One or more full-time technicians are needed to perform video control duties and to help the volunteers with technical concerns during studio shoots. When portable equipment is loaned out or when it is returned, a cable technician, together with the borrower, checks the equipment.

Lastly, we will enumerate the volunteer positions that are necessary to do a studio shoot. The volunteer users work at the center, but not for the center. To determine how many there should be, the number of people must be multiplied by the number of programs anticipated on the access channel.

In the studio, the center needs two camera operators, a floor manager, and the people referred to as "talent," in whatever numbers are necessary. In the control room, the producer, the director, and the switcher operator are required. These roles, under worse-case conditions, may all be performed by the same person. Also needed are the character generator operator and the audio control operator, who generally starts and stops the video recorders, in addition. As stated earlier, video control, via the camera control units, is handled by a cable technician since it requires experience in the use of the video wave form monitor and the vector scope. In many access control rooms, the access facilitator or a trained and skilled volunteer may be capable of this task.

C H A P T E R   5

SELECTING ACCESS CENTER EQUIPMENT

The selection of all of the electronic equipment, the lighting instruments, and the mechanical equipment that comprises the access television production complex is a very important one-time consideration, given the paucity of funds usually allotted for this purpose. There is also the further consideration that the gear must last for an indefinite period, with little likelihood of replacement until it virtually falls apart. This should suggest to the reader that the choice of equipment, and its replacement with time, should be seriously addressed in the writing of the franchise agreement.

EQUIPMENT SELECTION METHODS

Where then does the manager of a new access center look when considering equipment purchase, whom does he ask, what does he buy? Two broad scenarios govern the possible responses to these questions:

1. The access manager is starting a center in a community with an already written cable franchise that includes no equipment assistance from the franchisee, but does include an unfunded public access channel.

2. The access manager is starting a center in a community where the cable franchise has not yet been awarded. Or, it has been awarded, and the franchise contains one or more public access channels, and capital assistance from the franchisee for the purchase of access equipment.

If the scenario is number one, we suggest that the manager find out what funding is available in the community, with the help of the access corporation board of directors. Then consider the purchase of Beta 1 consumer-level record/playback equipment and consumer-level color cameras to begin your operation. The picture quality produced with this type of equipment will look relatively bad on the screen compared with broadcast television and the premium movies on the other channels, but it is a beginning. With successful and innovative programming, an upward move to industrial-level equipment will hopefully be possible in the future.

If scenario number two is the case, the rest of this chapter offers guidance for your equipment choice considerations.

The equipment—the cameras, their zoom lenses and tripods, the video recorders and editing controllers, the picture monitors, switchers, generators, audio control board and mikes, the other peripheral equipment—must all meet three basic criteria. They must be easy to operate and user-friendly; they must be very ruggedly constructed, yet lightweight if portable use is contemplated; and their electronics must be modular for maximum ease of maintenance. Modular construction in this context means plug-in printed circuit boards that can be easily transposed to insure minimal equipment down-time.

All of these constraints are necessary because the access operations gear will be used primarily by inexperienced, nontechnical people and repaired by technicians who are trained primarily to replace printed circuit boards, rather than to repair the faulty components on the boards. And even prudent neophyte volunteer cameramen will be inexperienced in their use of the equipment, which in itself can be cause for accidents.

Adding to this mild dilemma is the concern of the equipment purchaser, who lives in an environment where newer, better, and less expensive equipment arrives daily in the marketplace. He thinks as follows: "If I wait, will I get better equipment for less money?" That is a valid concern, and the trepidation is very normal. However, the concern must be discarded cleanly when it is time to purchase, because otherwise one could wait forever. While the cutting edge of the state of the art continued to become sharper, the access center would be without equipment—still waiting for the newer, the better, the cheaper.

The best advice that we can give to a prospective equipment purchaser is to talk to people actually doing access. Ask and write down their opinions on what equipment works best for them, and what equipment does not, and what would they buy now if they had it to do over again. The purchaser is urged to talk to many such people to get a broad spectrum of opinion from different access centers. Inevitably, the consensus of their opinions will be the recommendation of industrial-level electronics over broadcast equipment, which is many times more expensive, or over consumer-level equipment, which is much less expensive but suffers seriously in terms of quality. Consumer-level equipment is improving, however, and may be a viable choice in the near future.

Another more interesting facet of the consensus is that your group of respondents will probably limit their range of choices to perhaps five manufacturers for cameras, three or four for VTRs (VCRs), two for audio equipment, and even fewer for other peripheral equipment. The task of choosing, therefore, should not be very difficult, providing that reasonable adequate funding is available for equipment purchase.

It will also be valuable, in the course of choosing, to read the specification sheets and brochures of the recommended equipment manufacturers. They will gladly provide this information, but the reader should bear in mind that brochures, even those containing spec sheets, may include some puffery and are self-serving. In this book, we have chosen for the most part not to attempt to highlight specific models or manufacturers. When beginning the search for equipment, the reader is recommended to consult *Television Operations Handbook* by Robert S. Oringel (Focal Press), which gives detailed descriptions of the various makes and models of equipment.

A DESCRIPTION OF PRODUCTION EQUIPMENT _____

Studio Control Room Equipment

One of the most critical pieces of equipment in access is the portable video camera and its zoom lens (see Figure 5.1). Because the camera may be dropped or otherwise mishandled during location shoots or tipped over or harshly handled on an unlocked tripod in the studio, we stress the importance of equipment ruggedness. The risks to the camera also motivate our suggestion that the access center protect itself against equipment abuse by assigning binding monetary responsibility to the user, perhaps through an intermediary insurance policy. We discuss insurance policies in Chapter 6.

A major quality concern for a camera is usually its "resolution," or lines per complete picture. You may know that American television follows the NTSC (National Television Standards Committee) standard of a maximum of 525 lines.

Practical technical considerations, however, limit the normal industrial-grade video tape recorder—into which the camera will feed video signal—to a pickup and playback resolution approaching merely 300 lines (see Figure 5.2). The purchaser could therefore buy a camera with 600 lines' resolution, but when its video output is fed to a video tape recorder, the scene would be recorded at about 300 lines. This is a case of the chain and its weakest link, and the weak link here is the industrial-grade VCR. The admonition, then, is not to purchase an item of greater quality, at greater expense, than the whole system can use.

FIGURE 5.1A A video camera. Courtesy of Ikegami.

FIGURE 5.1B A pan-tilt head. Courtesy of Quick-Set.

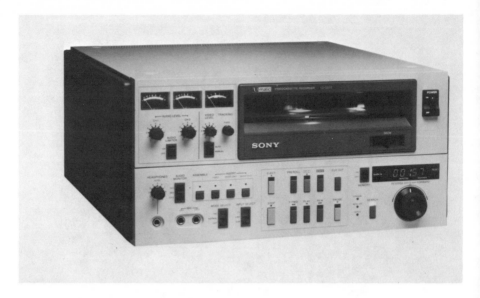

FIGURE 5.2 A studio video tape recorder (VCR). Courtesy of Sony.

A second camera-related concern is its color rendition. Are reds really red, or a dark shade of pink? This question may only be answered by a side-by-side comparison with results from a camera whose quality is known.

Apart from the camera, certain items of television gear just naturally receive more than their share of wear and tear. Camera cables and their connectors and audio cables and their connectors are perfect examples (see Figure 5.3). In this category, you are urged to buy the best that is made! And having done that, you may rest assured that even the best will not suffice. Plan a program early on to train people to use soldering equipment, and to learn to replace and check cable connectors. It is an easy task if it is demystified. Studies of commercial broadcast facilities have shown that the weakest television subsystem, in terms of failure during a program, is its audio (microphone) cabling. Your access center will never have too many cables available, and cable down-time will jeopardize programming.

Small items such as earphones and microphones tend to receive rough treatment and disappear through loss and theft. (See Figure 5.4.) It is advisable, therefore, to highlight these items during the checkout and subsequent checkin of loaned portable equipment. However, the access center must not buy cheap mikes and earphones, on the assumption that they will be lost and require replacement. Without good mikes sound quality

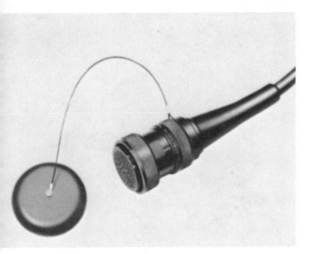

FIGURE 5.3A Camera connector.
Courtesy of RCA.

FIGURE 5.3B Video connectors.
Courtesy of Amphenol.

FIGURE 5.3C Audio connectors. Courtesy of Switchcraft.

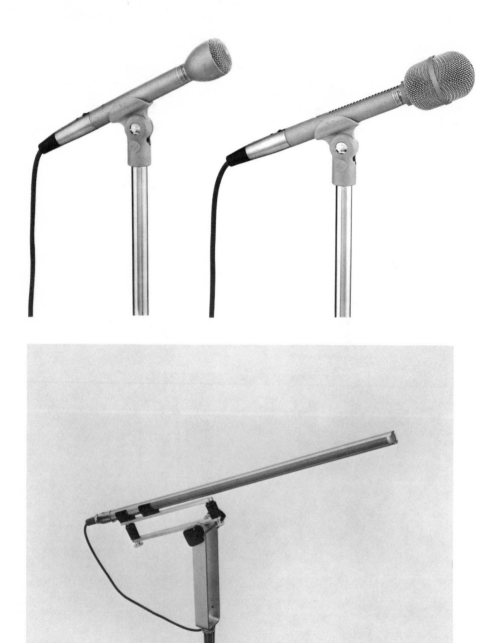

FIGURE 5.4 Microphones and connectors. Courtesy of Electro-Voice and Sennheiser.

will assuredly be bad, and sound is fully half of the communicating medium. If it is not already, this fact will become eminently apparent on the very first occasion that you lose sound during a program.

Poor-quality earphones provide poor-quality audio to the ears of the recordist, and this can cause serious audio judgment errors. Earphone cables and connectors should also be checked regularly for fraying cables and broken plugs.

Let us now discuss in detail the equipment that your access center will need. Since budgets vary so widely, we can only speak in terms of the nearly ideal. The ideal expressed here can be shifted upwards or downwards somewhat, depending upon the available capital, but all of the video equipment should be color gear, except perhaps a few of the video monitors, which may be black and white.

We look first at color camera equipment. Two types of cameras are required, heavier studio cameras and lighter-weight portables. Optimally, all of the lenses should be 10:1 zoom lenses. The studio cameras recommended at this writing are "three-tube" cameras, with all that they imply for the optics systems and electronics of all three pickup tubes. Portable cameras may have three-tube systems, but there are lighter-weight one-tube cameras. In fact, even lighter-weight industrial-level CCD (no tube) cameras are made, which are more than adequate and measurably almost as good as the three-tube systems. Access volunteers include people who would be prevented from participating in an outdoor shoot by the weight of a three-tube camera on their shoulder.

Portable cameras should have eyepiece viewfinders, and studio cameras should have the $3\frac{1}{2}$- to 5-inch camera-top mounted viewfinders (see Figure 5.5). Studio cameras should have remote control cables attached to controls terminating and mounted on the pan-tilt handles to adjust focus and perform zooms.

The pan-tilt handles of a studio camera are mounted to the pan-tilt head, which is mounted atop a tripod. A pan-tilt head will have both a "locking" and a "drag" feature in each of the panning and tilt modes. Make sure that the pan-tilt heads, as well as the tripods, that you get are sturdy and rugged. The camera's life literally depends upon it. Studio tripods (often called "trollies") have three wheels, usually with wrap-around bumpers to prevent wheel-to-camera-cable problems (see Figure 5.6). Broadcast television uses heavy pedestals borrowed from motion picture work to mount studio cameras, but they are too expensive to be covered by an access budget. Buy studio tripods with a low center of gravity. Pan-tilt heads that can be cranked up and down vertically are most helpful, too.

Also in the studio one finds a nineteen-inch color monitor on a stand, often with an audio loudspeaker included. There also should be micro-

FIGURE 5.5 Viewfinders. Courtesy of JVC.

phone cables and connectors for at least six mikes, and at least that many clip-on mikes should be available, too.

The studio should have a set of lighting fixtures. The complexity of these fixtures and the overall wattage required depends on the size in cubic feet of the studio room and the height of its ceiling. Formulas supplied by the lighting instrument manufacturers will determine the necessary wattage.

In the access studio control room, one finds a long desk or operations console. At, on, or in this desk is the rest of the control equipment necessary to do a production.

We will look at the desk from right to left, and describe its equipment (see Figure 5.7). Traditionally, the audio operator sits on the right end of the desk. In front of her is the audio mixer, a consolette with at least eight

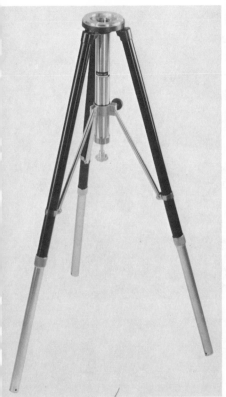

FIGURE 5.6 Tripod and trollie. Courtesy of Quick-Set.

available audio inputs, (the six mikes, an audio tape input, and a record turntable input), a good-quality audio speaker, a nine-inch video monitor that may be black and white, a record turntable, and an audio cartridge recorder/playback. Cassette and open reel audio playback machines may be available here as well.

To the left of the audio operator is the character generator and its video monitor. This must be a color monitor, preferably twelve inches or larger, because the character generator should have color generating facilities, including various color backgrounds and color highlighting of alphanumerics, which must be previewed. The character generator should have as many varying fonts of alphanumerics and as many pages of memory as the budget allows.

To the left of the character generator operator is the production switcher. In front of its operator there must be a nine- or twelve-inch monitor for each operable video input to the switcher. Each camera in the stu-

FIGURE 5.7 Control room console desks with equipment. Courtesy of the NFLCP photo files.

dio is an input, as are the VCR playback machines and video lines from other locations. These monitors may be black and white. The switcher operator must also have a preview monitor and a program monitor; both should be nineteen inches and color. In close proximity to the switcher should be two video cassette recorders, one for playback and one for recording, and in front of them should be a videotape editing controller for on-line editing. Each VCR should have its own nine-inch color monitor.

The switcher operator in access cable television is often the program's director, as well. If not, the director sits behind the switcher. The director should have an audio loudspeaker so that he can hear program audio, and he should wear an intercom headset with a single earphone for communication with the studio cameramen and floor manager.

To the left of the switcher are the three engineering monitors: a twelve- or nineteen-inch switchable input color monitor, a vectorscope, and a wave form monitor. Because these last two video measurement oscilloscopes are so critical to the operation of all of the other video gear, we recommend that you go first class and buy Tektronix equipment, the Rolls Royce of the industry.

Other pieces of equipment are on or are a part of the desk: one or more time base correctors on the VCR outputs, a camera control unit (CCU) for each studio camera, a video patch bay and an audio patch bay, and video and audio distribution amplifiers.

Portable Equipment

Portable on-location equipment is of two varieties. The first is the porta-pak, a portable camera and VCR, together with ancillary equipment like batteries, cables, carry cases, and tripod. The second is the remote van, which may either be a vehicle to carry porta-pak equipment, or a motor home type of vehicle permanently outfitted with all of the equipment found in a studio control room built in, plus storage room for a sufficient number of cameras and reels of camera cable to do an extensive outdoor shoot of a complex sports event.

Editing Equipment

In addition to the editing equipment found in the studio control room, most access centers include one or more editing rooms or suites in which to do video tape editing. One often hears the terms "on-line" and "off-line" editing. In broadcast television, off-line editing refers to a videotape work print with "burned-in" timecode that is viewed by a producer in his office on a VTR playback and monitor. He makes careful notes for edit "ins" and edit "outs," using the timecode as his reference. These notes are then given to a videotape editor who does the actual editing on an "on-line" facility with full editing capability.

In access, on-line usually refers to editing done in the studio control room, while off-line refers to editing done in a separate editing suite. The editing room usually includes a desk with a recording VCR, a playback VCR, and an editing controller (see Figure 5.8). There should be a twelve-inch color monitor for each VCR and an audio loudspeaker as well. Adding audio sources—like an audio cassette or an open reel audio tape playback, or a character generator—to the editing suite makes it almost equal to the control room in terms of editing capabilities, so that the terms on-line and off-line have little practical meaning.

Before leaving the topic of the editing room, we must consider one more item: videotape. Although it is found in the operating budget rather than in the capital budget, videotape is nevertheless a vital part of the operations system. There is an ongoing need for and expense of videotape, which must be available at the center in large amounts—in twenty-minute, thirty-minute, and sixty-minute cassettes.

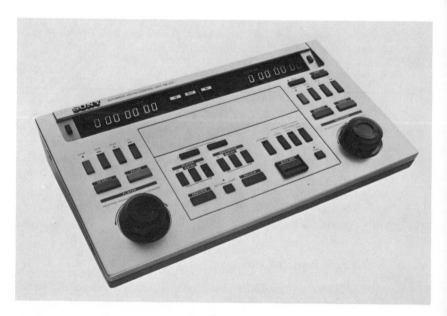

FIGURE 5.8   Editing controller. Courtesy of Sony.

TRANSMISSION EQUIPMENT

Having described the equipment necessary to make programs, we conclude with a description of the equipment necessary to cablecast programming on the cable system's access channel.

The most basic requirement is a single VCR playback capable of playing back an hour-long cassette, with its output connected to a modulator feeding the channel. However, to prevent the access audience from watching a blank channel as one tape is removed at its end and replaced by an uncued second tape, the playback system must be sophisticated enough to include a second VCR playback, a simple vertical interval switcher with "audio follow video," a video and audio monitor that is switchable between the access channel and each of the playbacks for cueing, and a character generator to feed the channel when no access programming is on line. For the rare access centers with a great deal of capital, there are automated videotape playback systems available that, once loaded and cued, play the day's programming in sequence, without requiring a staffer or volunteer to be on duty whenever access programming is on the channel.

LAYING THE GROUNDWORK FOR A SUCCESSFUL ACCESS CENTER

The successful center—in contrast to the center that always struggles to keep in business—from the outset will have taken care of at least five major aspects of operation. These are budgeting and financing, insurance, record keeping, copyright clearance, and designing forms.

BUDGETING AND FINANCING

Budgeting can be described as figuring out how much money will be needed for a year's operating and capital (equipment) expenses. Budgeting entails close scrutiny of all of the costs of operating an access center: the rent, electricity, staff salaries, production costs, equipment upkeep costs, insurance policies, office supplies, videotapes, expenditures for new or replacement gear, and all of the other myriad financial requirements during a calendar or fiscal year. Once an accurate cost is assigned to each item, the budgeter totals the expenses and adds in a percentage for emergencies, inflation, and unexpected expenses. The final budgeting task is to encourage everyone concerned to make a heroic effort to keep within the figure that has been calculated and financed.

Financing can be described as the methods or strategies used to obtain the funding for the total budgeted cost figure—and perhaps a bit more. Such money may come from the cable company or from use of the access facilities or access tapes by outside organizations (if those facilities are owned by the access corporation, rather than by the cable company). Use of facilities by outside organizations should not, however, compromise access policies. Finally, money may come from fund raising by the access organization. Besides earning necessary money, fund raising, if done using the television facility, can be a great learning experience and a lot of fun for the access center staff and volunteers. Fund raising can involve telethons, creative games like "TV Pursuits," or any other constructive idea using the TV medium. This is done both to entertain and to convince viewers to donate to help defray the costs of public access television. The Public Broadcasting Service is an excellent example of how well fund raising can work.

INSURANCE POLICIES

The insurance policies that are essential to protect the access corporation and its staff—both paid and volunteer—are discussed below.

- Board of directors' liability. This policy protects the members and directors of the access corporation from any personal legal liability that may arise from excesses or deficiencies in the operation of the corporation or of the access center. Suppose, for example, that someone was injured by falling over a mike cable or camera cable and that the person sued the access company for damages and won. If the company were not covered by sufficient liability insurance, individual corporate directors might be legally liable.

- Fire and theft. This policy protects the access center premises and equipment from hazards like fire and theft.

- Health and worker's compensation. This policy provides health insurance and worker's compensation for employees. Most states require worker's compensation insurance for employees.

- Bonding. This policy provides bonding for the treasurer and any other officer who handles large sums of corporate money.

- Equipment user's liability. This policy protects equipment users against full liability for accidental loss or damage to access center equipment, and limits their liability to a deductible of perhaps $500. Please note that we specify accidental loss or damage, as the insurer most certainly will. Damage or loss that can be attributed directly to the negligence of the user is usually not insurable.

- Cablecaster's liability. This policy protects the access company monetarily if it is named as a defendant in a defamation suit—that is, libel or slander; invasion of privacy, commercial appropriation, false light, intrusion, or disclosure of embarrassing facts; infringement of copyright or trademark; plagiarism, or piracy of ideas; and errors of omission or misstatement or erroneous information. This insurance covers program content, and if it is not in place, a lawsuit could cost the access corporation more than it could pay.

We earnestly suggest that the access manager look long, hard, and carefully for an independent insurance agent who will spend the time and do the research necessary to come up with a policy plan to cover the liabilities listed previously. At the same time, the agent should keep the cost of such insurance at a reasonable level. In addition, the manager or the agent should look for commercial package policies that will cover several if not all of the items under one policy, because package policies tend to be less expensive than individual coverages.

The company should avoid purchasing policies from an insurance person who is a member of the access organization's board of directors. Even though this person might know just what the center's needs are and might be able to furnish the best price, the appearance of conflict of interests overrides any possible benefits or savings.

DEVELOPING RECORD KEEPING SYSTEMS

One need not dwell on the importance of record keeping, particularly in an environment as fluid as an access center, where volunteers come and go and where accountability for expensive items of equipment must be maintained.

The most practical suggestion that can be made is to start with a small computer as a record keeping device. A computer, which need not be expensive, with a word processing program, can double as the office typewriter, and can be less expensive than a good quality typewriter. The word processor can be used to prepare the inevitable center newsletter and to produce all of the correspondence that any office generates.

With a data base file manager program, the computer can keep a record of each center user, of budgetary items, of equipment checked out and checked in, and of any and every record keeping chore that is necessary to the center. Using a file management program, the computer can prepare three-by-five file cards for each of the center's members and users, including a cross-file listing members by either name or by operating skill. The computer can periodically print new cards that reflect new information on membership dues and training, generate membership lists for the company treasurer or whoever else needs them, and provide individual gummed labels for mailings to the membership or to any other subgroup— such as board members, producers, or camera operators—that is separately identified in the data base system.

Concern is occasionally expressed that files saved on magnetic media (such as floppy disks) can be destroyed by stray electrical or magnetic fields. Such losses can easily be prevented by making back-up copies of file disks and storing the back-ups away from the originals.

The computer is a versatile tool, and can be used creatively in many aspects of the center's activities—including the video operations.

LEGAL CONCERNS OVER COPYRIGHT CLEARANCE _____

Copyright Clearance

All creative expressions, such as play scripts and music used as parts of television programs, are copyrighted at the instant of creation by their authors or composers. Copyright is authorized by Article 1, Section 8 of the US Constitution. A copyright may or may not be registered with the Copyright Office of the Library of Congress. Copyright of a work means that the author or his agent have and retain exclusive rights over the use of their properties for a specific period of time. Further—and most important—the copyright holders must grant permission and possibly be paid fees before their properties may be used in a television program.

Obtaining such permission is called copyright clearance. Using copyrighted material without clearance places the access center, and particularly the program producer, in an indefensible position if the author or composer decides to sue for damages. However, small portions of music or writings may be employed without clearance under the "fair use doctrine."

As we have stated, copyright protection starts automatically at the instant of creation. No formal procedures are necessary as long as each copy of the work contains a copyright notice. The copyright notice for "visibly perceptible" copies should contain the following three elements: (1) the copyright symbol, ©, the word "Copyright," or the abbreviation "Copr"; (2) the year of first publication of the work; and (3) the name of "the owner of copyright in the work."

Any work copyrighted after January 1, 1978, is automatically protected from the moment of its creation to the end of the author's life, plus fifty years. If there is joint copyright with two or more authors, the copyright term lasts for fifty years after the last surviving author's death. Copyright protection extends to "all original works of authorship, fixed in any tangible medium of expression." That includes music, drama, motion pictures, audio-visual works, sound recordings, pictorial works, graphics, sculpture, videotapes, computer programs, and any other form of expression that can be fixed in tangible form. The legal formality of registering the copyright with the Library of Congress is done by completing an application and sending it with a $10 fee and two copies of the work to the Copyright Office.

The copyright holder is authorized to reproduce the work, to do a derivative work, to distribute the work, and to perform the work publicly.

At copyright termination, the property is said to be in the "public domain" and may be used without further clearance.

Many currently popular playwrights will not permit their major works to be performed by local community theater groups specifically to be shown on cable channels. This is because the fee charged for the "grand rights" to the work is normally for one performance, or one small group of performances to a small theater audience and to no other audience.

From the author's or composer's point of view, video performances, such as those on live cable television, will be videotaped and repeated at the discretion of the access center, without additional payment of fee. Thus the community theater group that wishes to perform on an access channel—either from their own theater or in the access studio—must seek out local and less well known playwrights. These local authors are often very grateful for the opportunity to expose their works on television. Local authors and composers can be found in virtually every community and often they will gladly charge little or nothing for clearance to perform their works.

The Fair Use Doctrine

The fair use rule permits small portions of a copyrighted work to be used for "socially meritorious" purposes. The doctrine is a part of the new copyright law, which was rewritten by the US Congress in 1976. The law balances the limited monopoly given to copyright holders with the legitimate needs of the public at large. Its premise is that there are limited instances in which copyright material may be used without permission of, or payment to, the copyright owner, provided that the use is reasonable and does not unduly harm the rights of the owner.

Section 107 of the Copyright Act states four criteria for whether a particular use is fair:

1. The purpose and character of the use, including whether it is for commercial purposes or for nonprofit or educational purposes.
2. The nature of the copyrighted work.
3. The amount and substantiality of the portion used in relation to the copyrighted work as a whole.
4. The effect of the use upon the potential market of, or the value of, the copyrighted work.

The law further adds that these criteria are not the only ones that may be considered by a court in determining fair use. Additional factors that may be considered include whether the user was acting in good faith and whether any deception has been involved.

A persistent and mistaken belief that is held by the public, and that is at least as old as the broadcasting business, is that a certain number of words or musical notes can be used without permission. *No such standard has been endorsed by the Congress or by the courts.*

The reader can easily see that fair use doctrine is very loose indeed, and subject to all sorts of court interpretations. The best advice that we can give under these circumstances is to consult with competent legal counsel before using copyright materials.

DEVELOPING ACCESS CENTER FORMS

The use of forms, to be completed by the access center or access equipment users, is a structuring feature that is very necessary to access center management. Forms keep the center users aware of the need for order and provide them with a sense that their formal promises to the center must be kept. Completed forms also mean that there is a written record of the center user's training and programming decisions.

Among the necessary forms are: application for membership in the access corporation; application for access training; request for and contract for portable equipment loan; request for studio time; request for editing time; request for time on the access channel; copyright clearance request; cameo appearance request; and talent release. The forms named are self-explanatory, except for the last two. A cameo appearance request is made by a community member who wishes to appear briefly on the access channel, perhaps to read a poem or offer a birthday wish to someone, without being involved in any other way in access. A talent release form is signed by anyone appearing on the channel, who agrees to hold the access corporation, the cable operator, and the community blameless for any legal action that occurs because of the appearance.

Some of these forms are supplied by the cable company, but cable company forms are often hard to understand, as though written by and for litigants, and are often issued merely for the sake of having a form, rather than for presenting necessary information.

An access center form should ask what it wants to know in clear and precise language, and it should also state its purpose clearly. It should leave sufficient space following questions for the answers and statements. The form should be written in clear English and should be understandable to anyone who reads it.

Samples of access forms are available in a packet prepared by the NFLCP. Please see Appendix B at the end of the book for the NFLCP's address.

TRAINING CURRICULA
AND COURSE CONTENT

A SYSTEMATIC APPROACH TO USER EDUCATION

We begin this chapter by stating the obvious: Training is necessary because public access centers do not have a cadre of trained technicians and program producers to do programs.

The training of access center users is important to the access manager because the training procedures set the tone for the relationship that will develop between the user and the center. The motivation to work in access has to be developed in members of the community and it must be continuously sustained if people are to continue doing programs. Further, in our experience, an individual's interest in access will often wane after a time, so that the user must be replaced and the replacement must be trained and motivated. Strategies for user replacement are covered in Chapter 8.

For the training process, a systematic approach to user education should be developed at the center. We recommend that the initial training progresses in three stages: from location equipment training, to studio production and equipment training, to videotape editing and postproduction training. Producing and directing skills can be interwoven into each of the three levels. This is a logical progression, and although it should not be etched in stone, many successful access centers seem to proceed in this way.

PHASES OF TRAINING

Location Production Facilities Training

Location training teaches the use of the portable camera and the portable VCR, which together with some peripheral devices such as a microphone and some cables and powering units, are called a porta-pak. These two devices are taught apart from—and usually out of sight of—the possibly confusing array of equipment in the control room. People who are new to access equipment should be exposed to it in manageable doses.

About eight hours of training, with an instructor and a class of about half-a-dozen students, provides the rudimentary essentials and a bit of hands-on experience for each class member. In many training situations, the group will conclude its efforts by producing a short program to be cablecast on the access channel. This work provides an immediate and ideally positive first production experience.

The basic initial training should be followed by a test, either practical or on paper, to indicate that the individual has mastered the basics. Passing the test leads to the award of a certification card. This coveted certification qualifies the graduate to borrow equipment and to further enhance her equipment skills by beginning to make simple location television programs.

Studio Production Facilities Training

The second level of training involves learning how to operate most of the facilities in a studio and control room. This level expands the trainee's knowledge to include microphone use; the multitude of skills associated with the audio control board; the use of the character generator for the titles that open and close a program; the production switcher, which mixes video inputs and generates special video effects; and the VCRs that act as program input sources and program recorders. Most important, this second level of training also introduces the concepts of video production and direction.

Videotape Editing and Postproduction Training

The third training phase teaches the skills necessary to do videotape editing and provides the programmatic concepts necessary to use those skills for video postproduction.

More advanced or expanded training might include advanced field production; specialized lighting and audio seminars; basic stagecraft; making flats, drops, and other stage illusions; or advanced editing and post-production training.

CREATING THE CURRICULUM

From an educator's perspective it is always advisable to create a curriculum or standardized pattern in advance of training. Given the differences between trainers, the curriculum attempts to insure that trainees will all be exposed to much the same information and that they will receive about the same amount of experience in using the equipment.

A well-developed video curriculum should include required reading material on the use of the equipment and on skill techniques. Since video is the medium being taught, the curriculum should also include videotapes showing operational techniques and examples of good and bad programming. In addition, the course should have printed handouts that clearly diagram the ways in which equipment such as the camera and VCR are interconnected. Simple diagrams or line drawings should be used to display all of the switches and controls on the various kinds of equipment at the center (see Figure 7.1), and self-testing questionnaires should be issued to build user self-confidence in the trainees. Trainers should be urged to follow the curriculum as closely as possible.

A training curriculum may be thought of as being separated into three parts: design, implementation, and evaluation.

Designing the Curriculum

The first step in curriculum design is to determine who in the community the prospective students will be and what their training needs will be in terms of a television operations course. The ages of the students taking the course (which may range widely in the same class); the students' previous video experience and background or the lack of it; their individual reasons for taking the training; their particular problems or pressures (such as fitting the course into busy schedules, or possible physical handicaps, or individual job expectations) must all be considered in the design phase.

When assessing the trainee's needs, the designer must ask how they will use the skills obtained through the training. Some will go on, perhaps, to jobs in broadcasting; some will use their new knowledge for a home video hobby. What about their individual levels of commitment to access programming? What are their interests in access? Do they want to operate production equipment? Do they want to produce access programs?

Having considered the student population, the curriculum designer must then focus on the goals of the curriculum. Are the overall or long-range goals merely to impart information, or are they to do more? Will the goals that are built into the curriculum create involvement in access? Will they foster an awareness of access production's relationship to the First Amendment rights of citizens? Will training goals provide a successful first experience in access? Will they imbue students with the access philosophy, and teach students to work as a team? The access manager's answers to these questions will create the parameters for the long-range goals of the curriculum.

There are also some immediate goals to be considered. These are usually stated as objectives or skills that the student will reach within a given time—perhaps a single class period or a specific course. Some of these objectives are: to operate a given piece of equipment (a camera, a VCR, or a production switcher); to write a script in an approved format with full directions included for video, audio, and talent; to describe the ways in which access programs are distributed within a community; to identify the

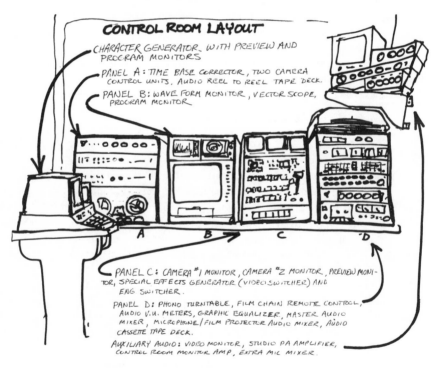

FIGURE 7.1 Simple line drawings of access equipment devices, switches, and controls. Courtesy of the NFLCP files.

AUDIO MIXERS (PANEL "D")

HEADPHONE JACK: BETTER TO USE THE JACK UP ABOVE IN THE METER PANEL SO YOU CAN HEAR THE AUDIO AFTER IT HAS GONE THROUGH THE EQUALIZER.

HEADPHONE VOLUME CONTROL: NEED WE SAY MORE?

ROW OF EIGHT "PAN POTS": SENDS THE AUDIO SIGNALS FROM THE CORRESPONDING VOLUME CONTROL TO EITHER CHANNEL #1 OR #2 OF THE RECORDING VTR. TURN LEFT TO SEND TO CHANNEL #1, RIGHT FOR CH. #2, MIDDLE FOR BOTH.

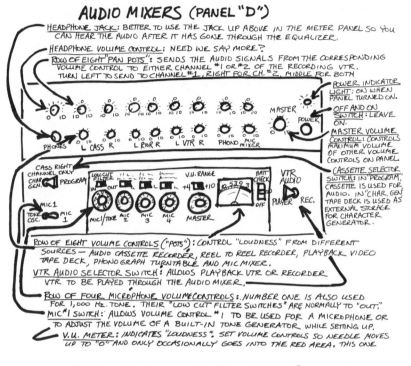

POWER INDICATOR LIGHT: ON WHEN PANEL TURNED ON.

OFF AND ON SWITCH: LEAVE ON.

MASTER VOLUME CONTROL: CONTROLS MAXIMUM VOLUME OF OTHER VOLUME CONTROLS ON PANEL.

CASSETTE SELECTOR SWITCH: IN "PROGRAM," CASSETTE IS USED FOR AUDIO. IN "CHAR. GEN." TAPE DECK IS USED AS EXTERNAL STORAGE FOR CHARACTER GENERATOR.

ROW OF EIGHT VOLUME CONTROLS ("POTS"): CONTROL "LOUDNESS" FROM DIFFERENT SOURCES — AUDIO CASSETTE RECORDER, REEL TO REEL RECORDER, PLAYBACK VIDEO TAPE DECK, PHONOGRAPH TURNTABLE AND MIC MIXER.

VTR AUDIO SELECTOR SWITCH: ALLOWS PLAYBACK VTR OR RECORDER VTR TO BE PLAYED THROUGH THE AUDIO MIXER.

ROW OF FOUR MICROPHONE VOLUME CONTROLS: NUMBER ONE IS ALSO USED FOR 1,000 Hz. TONE. THEIR "LOW CUT FILTER SWITCHES" ARE NORMALLY TO "OUT."

MIC #1 SWITCH: ALLOWS VOLUME CONTROL #1 TO BE USED FOR A MICROPHONE OR TO ADJUST THE VOLUME OF A BUILT-IN TONE GENERATOR WHILE SETTING UP.

V.U. METER: INDICATES "LOUDNESS". SET VOLUME CONTROLS SO NEEDLE MOVES UP TO "O" AND ONLY OCCASIONALLY GOES INTO THE RED AREA. THIS ONE

AUDIO METERS AND EQUALIZER (PANEL "D")

HEADPHONE JACK: PLUG THEM IN HERE TO HEAR THE AUDIO WITHOUT DISTRACTION.

HEADPHONE VOLUME CONTROL: ADJUSTS VOLUME TO THE HEADPHONES ONLY.

TWO PROGRAM V.U. METERS: INDICATES LEVELS OF SOUNDS LEAVING THE AUDIO MIXER ON THE LEFT (CHANNEL #1) AND RIGHT (#2) CHANNELS.

TWO REC VTR. V.U. METERS: SHOWS LEVELS ACTUALLY GOING THROUGH RECORDING VIDEO TAPE RECORDER. THIS IS SYSTEMS FINAL V.U. METER.

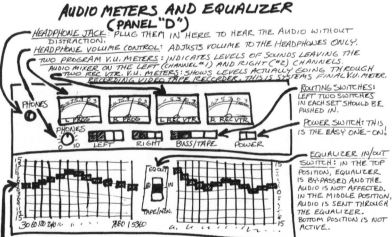

ROUTING SWITCHES: LEFT TWO SWITCHES IN EACH SET SHOULD BE PUSHED IN.

POWER SWITCH: THIS IS THE EASY ONE-ON!

EQUALIZER IN/OUT SWITCH: IN THE TOP POSITION, EQUALIZER IS BY-PASSED AND THE AUDIO IS NOT AFFECTED. IN THE MIDDLE POSITION, AUDIO IS SENT THROUGH THE EQUALIZER. BOTTOM POSITION IS NOT ACTIVE.

EQUALIZER SLIDER POTS: INCREASES OR REDUCES THE AUDIO LEVEL OF DIFFERENT FREQUENCY RANGES. THE SCALE AT THE BOTTOM INDICATES THE FREQUENCY, THE LOWER THE NUMBER, THE LOWER THE PITCH. THE SCALE AT THE SIDE INDICATES THE DEGREE OF INCREASE OR DECREASE, WITH THE CENTER POSITION BEING NEUTRAL, NO CHANGE. LOWER HALF IS DECREASING THE VOLUME, MOVING THE LEVER UP INCREASES THE VOLUME. THE LEFT SIDE AFFECTS CHANNEL #1 GOING TO THE VIDEO RECORDER, RIGHT SIDE AFFECTS CHANNEL #2.

personnel necessary for a television production and to describe each of their program responsibilities; to rewrite a print media message so that it is suitable for video presentation.

Implementing the Curriculum

A curriculum is implemented by using fairly standardized teaching methods. Some of the typical methods to help students achieve the goals that have been established for the class are: lectures by an instructor; demonstrations of equipment use or of production skills; class discussion; supervised hands-on practice of specific tasks; hands-on drill of specific tasks; production with the instructor in the most responsible position; production with the instructor in an advisory position only; production with no guidance as a test, followed by a critique.

The classic steps used in the teaching-to-learning process are: introduction of the subject, reinforcement of the key concepts, practice, performance, and critique of the results.

We advise instructors to listen carefully to their students' comments; these comments will provide valuable feedback on whether the material in the lesson is reaching them. Always reinforce the key concepts of the lesson in the clearest and best way. Whenever possible, let the student participate fully in an operational lesson. Finally, keep it simple and keep it enjoyable.

Evaluating the Curriculum

The evaluation process concerns itself with a single feature: success. How successful were the students, the teacher, and the course?

First of all, a curriculum evaluation program needs to develop ways to determine whether the students have accomplished their immediate goals. This is done by having students respond to direct oral or written questions, or by having them demonstrate their new-found abilities either individually or in groups.

Developing ways to determine whether the students have accomplished overall, long-term goals may be more difficult, since success may have to be measured over time. As an example, measuring access production attitude may be done through regular surveys of access crews, or by keeping track of the return rate of access producers, or even by an occasional survey of people who have left access production to ascertain why they left.

The success of the instructor is in large part measured by the success of the students. However, in some instances, students learn in spite of the teacher. At the end of a course, a brief survey of the students may be useful to check their perception of their teacher's instructional abilities.

How successful was the course? Did it fulfill the students' needs? Was the material relevant or was it out of date? The same end-of-course survey that rates the instructor should include questions on these factors.

Did the course fulfill the access center's needs in terms of providing new users? Did the fledgling access producers gain the skills needed to produce access programs? The training given at the center is, after all, for the sole purpose of facilitating access programming.

Course emphasis should be flexible enough to be changed to remedy problems that become evident during training or during program production. For instance, if new access crew members seem uncooperative, then the course might be changed to place more emphasis on developing crew teamwork.

COURSE CONTENT IN ACCESS TRAINING

Aside from simply learning to operate equipment, video production entails many creative and aesthetic considerations. We will examine a few of these considerations, and describe some of the material that we think should be included in the course content.

Planning an Access Program

To plan the program, start with a session that focuses on the basic questions asked in a radio or television interview: who, what, where, why, when, and how. Who will do what on the show, and how will the program responsibilities be divided? What is the program designed to say? Where is the program going to take place? Why do the program? When will it be cablecast, and is timeliness an issue? How will the program advance from an idea to reality?

Who. Teamwork is vital in video production. Everyone involved should have a clear picture of what their individual responsibilities are. For a production to be a success, roles of producer, director, camera operators, audio tech, lighting tech, production switcher operator, character generator operator, video tech, the talent, and others have to form a totally harmonious group. These roles are spelled out in detail in Chapter 12.

What. Every program has a message, even if it is sometimes obscure. The clearer the program's message—in terms of its purpose, focus, and objectives—the clearer it will be to its audience. Programs that try to persuade and attempt to influence viewers to think or act in a particular manner use "parent" or authority figures, such as government leaders, in formats like panel discussions, editorial statements, and question-and-answer shows.

Programs that try to entertain provide comedians, plays, music, and dance. Programs that are instructional or informative use demonstrations of techniques, classroom instruction, or slow-motion photography. Documentary programs show documents, give oral histories, and record history.

Where.   If the program is to be done in the studio, then, clearly, the scene can be no larger than the studio. For a location shoot, factors such as clearances, A/C power locations, light placement, and audio considerations must be dealt with. These factors are addressed individually in detail elsewhere.

Why.   The motivation for a program has to be some type of desire to communicate. *Co*-mmunication is a two-way process. In the audio-visual media, communication involves the intent of the communicator and the receptivity of the viewer. If the program has effectively transmitted a message to a receptive viewer, then some noticeable change will have occurred in the state of the viewer. To plan that change in advance is part of the program, part of the message. In his planning, the program planner places himself in the viewer's shoes and considers the viewer's response.

When.   Timeliness is the difference between a program that is news and one that is history. Words like "tonight," which date a program, detract from its use as a repeat.

How.   Advancing from the germ of an idea to the reality of a program on videotape involves a number of factors. These factors must focus sharply on the purpose of the program. The program has a subject, which has to be given a "treatment" in the program. A particular point of view, or angle, or slant must be used to give the viewer a sense or an experience of the subject. The program content should be creatively presented and should be in a script that describes the program flow in both audio and video. The program should have a story line, or plot, that moves from a beginning, through a middle, to an end. The plot should include situations, tensions, characters, and conflicts with which the viewer can identify and be concerned.

The program has to have a format or method of presentation. There are not many formats: one person speaking to the camera, a group of people—two or more—speaking to each other and seemingly ignoring the camera, or a group of people acting, singing, or otherwise cavorting on a stage. Each of these formats has innumerable variations, including the addition of external music, external video, props, sets, visuals such as slides or graphs, or combinations of these elements. What makes one program different and perhaps more successful than another is the way in which the programmer blends the format variations and includes new ideas for variations.

For a program's success, its creator must think about all of the various elements that comprise the show. What will the viewer see? Who will talk to whom? What and how much will the cameras describe? What angles will the cameras use to provide their descriptions? What graphics will be included? Will graphics appear in more than the opening and closing credits? Will they be generated solely by the character generator, or will there be printed cards, a scroll, or slides? Will the show have a logo?

Program planning includes—but is not limited to—doing program research, finding locations, scheduling shooting dates, getting clearances, and arranging for talent, equipment, and crew.

Script Writing

A television script is the means by which the programmer's ideas for the program are translated into images that other people can understand and act on. The other people are the talent, the crew, and the access center director. The script provides them with the words, the movements if any, and the pictures. The script should include an outline of the program content, in either chronological order, in thematic sequence, or within the time frame of cause and effect. The usual method used in television scripting is to present pages that are divided into two or three vertical columns. The first column contains the video action, the second contains audio script, and the third column, if used, is for videotape timing. The video column sets the shot, describes the action, gives movement directions for both cameras and talent, describes the special effects to be used, and details the transition to the next shot. Concurrently, in the second column, the audio is scripted, and the third column breaks down the time of each sequence in seconds.

Often "storyboards" are used to illustrate what will occur on the screen. Storyboards, like comic strips, present the screen action in shot-by-shot sequence. Beneath each shot's picture box on the storyboard is its audio, its description, and its camera instructions in another, smaller box.

Video communication is not an end unto itself. As we have stated, it is a means to transfer and exchange ideas and information. Because information is transferred by audible and pictorial means, the script should always describe the visuals that correspond to the spoken material.

In addition, we should point out that visual focus only on what are called "talking heads" or static conversation between people is deadly to a program's interest. Regardless of how interesting or informative the subject matter under discussion is, it may become boring to the viewer if it is presented under static conditions. Several techniques may be used to enliven the program. Include visuals of the discussed subject. Write the script with emphasis on pictures and let the pictures tell the story wherever possible. Write "action," remembering that video is a moving medium. Don't forget to include video or audio transitions from one scene to another. Use words to supplement the pictures and action, and choose the right words for the target audience. Keep the words informal and conversational. Don't be afraid of pauses—they are normal in discussion because it takes time to turn thoughts and ideas into words. And as to words, it is best to use "we" instead of "I," and "you" instead of "he" or "she."

Avoid dating the written material with "good evening" or "tonight" if the program is to be cablecast more than once. Be aware of the viewer's attention span, and break the program into segments if the subject material exceeds fifteen minutes on a long program.

Camera Shooting

Camera shooting and operation are clearly different. A discussion of camera operation would include the use of all of the switches and controls associated with the camera and its zoom lens. Camera shooting involves using the whole camera to take pictures. And taking pictures with a video camera for a program is subdivided into two categories: shooting with a director and shooting script, and shooting on your own. The scripted and directed method is simplicity itself. You follow directions, for the most part, only varying from the shooting script when the camera is not on or when a shot of opportunity comes along. A shot of opportunity should be framed in the viewfinder and pointed out to the director.

When shooting on your own, shoot more than you will need, but not so much that you will need extensive editing later. Always edit in your head while shooting. Shoot long shots and close-ups, plenty of cutaways, and a variety of interesting angles. Watch for "matched action" shots, but don't shoot long pans or long zooms that will be impossible to cut into later. Constantly look for transition shots. Shoot plenty of nonmoving material. Don't overload yourself with excess equipment on a location shoot, and if something goes wrong technically, try to keep the camera and VCR rolling. Perhaps something on the tape will be salvageable later. Above all, whenever you and a camera are on a shoot, try to develop and use what will later become an innate production sense or intuition.

Learn to visualize what the viewer will see, and know the relationship between the camera—in this context, the eyes of the viewer—and what the camera sees. A long shot or a wide shot will set a scene, but will not convey intimacy between the viewer and the scene. A close-up or a tight shot creates visual intimacy between the viewer and whoever is on screen.

Height relationships are also important in this context. If the camera looks up at the talent, then the talent seems dominant over the viewer. Conversely, if the camera looks down at the talent, then the viewer feels dominant. Create powerful images with the camera. Use the video camera effectively by taking advantage of its ability to pan, tilt, dolly and truck, and zoom.

Seasoned camera operators always follow some basic rules before, during, and after each studio production. Before studio production, put on the headset and test the intercom. Unlock the pan and tilt mechanisms on the tripod. If the camera is turned on (fired up), uncap the lens, adjust the viewfinder, and rack through and check focus. If the operator leaves the camera, even momentarily, the pan and tilt must again be locked.

Immediately before the production, with the headset on, establish contact with the director, unlock the camera, and set the pan and tilt drag. During the show, preset the zoom at each new camera position. Zoom all the way in and focus. Know the reach of the camera cable before starting a dolly or truck movement. Always be alert to studio traffic. Where are the other cameras? The floor manager? The floor monitor?

Listen carefully to the director's instructions to all crew members so that you will be able to coordinate your shots with those of the other camera(s).

At the completion of production, when the wrap-up has been given, lock the camera, cap the lens, and push the camera to its storage position in the studio. Coil the camera cable and then help the other crew members strike the set.

Being Talent

If you are going to be videotaped for access television, maintain eye contact with the viewer, through the camera lens, to create a one-to-one relationship with the viewer. Try not to read directly from a script or cue cards, because it is very difficult to sound believable if you are not a professional actor or news reader. What is worse, the amateur tends to stare at the printed matter rather than looking at the viewer. Try looking at run-down sheets or outlines of the topic before the program, and then move them to the back of your mind.

Move freely when you are in front of the camera, rather than seeming to be frozen in one spot. When you are part of a two-shot or a three-shot, it is better video for the talent to move than the camera. Talent movement sustains viewer attention. Movement should be direct and smooth, and should be a response to some sort of motivation. Movement can be for emphasis or for transition. Swaying back and forth in a seat is bad movement and should be avoided because it tends to make the viewer seasick. Avoid playing with the microphone or its cable because the program doesn't need the additional noise that this causes.

Wear appropriate clothing for the program. Always avoid wearing coarse pinstripes, plaids, or herringbone patterns, which cause technical interference problems that are sometimes appropriately called herringboning. Colors can also be a problem: stark white can cause flaring, and red can influence the way that the camera interprets skin tones. In short, concentrate on mid-tone colors, solid fabrics, and avoid extremes in both light and dark colors.

Talent, like everyone else, must be motivated to perform well. Self-motivation includes taking the time to know what is going on in the production of the program. Talent should be encouraged to use personal life experiences as a basis for playing difficult dramatic scenes. Talent gets tired, and occasionally needs to be given a break when the action shifts to another scene or to a recorded video insert.

Location Shooting

Location shooting should be preceded by a site survey that contains a checklist of points to be considered. Start the checklist by drawing a floor plan of the location site. Locate everything—cameras, props, set, and talent—on the floor plan. The next concern, especially for an outdoor shoot, is available light. Determine the quality and amount of the light, and the sun's angle at the approximate hour of the shoot. Observe the surroundings carefully for potential problems like the shadows of trees, nearby buildings, or hills. If the available light will not be adequate, or if there is any hint that it might be inadequate, locate a nearby available power source for lighting instruments. Remember to check for circuit overload and proper grounding plugs; the average single circuit can handle only fifteen amps, which translates into about 1500 watts of lighting.

Look for shooting backgrounds and alternatives at the location. Consider bringing a roll of seamless paper for a neutral background.

Consider crowd control and automobile traffic control if the shoot location is a public area. Decide whether the local police force should be made aware of the shoot in advance and asked for assistance.

Security is always an important aspect of location shooting. Find a secure spot to unload equipment and park vehicles, select a safe area to leave equipment, and choose someone to watch the equipment. Identify any problems that might arise for vehicles carrying the crew, the talent, or the equipment. If anything needs to be purchased at the site, such as food, bring sufficient funds and alert the vendor of the size of the requirement.

Select a suitable location for meals and arrange for proper trash disposal afterwards—otherwise, expect never to be allowed to use that site for future shoots.

Samples of training curriculum material are available from NFLCP. See Appendix B at the end of this book for the NFLCP's address.

COMMUNITY OUTREACH, PUBLIC RELATIONS, AND ACCESS RELATIONSHIPS

OUTREACH

Outreach, in the public access sense of the word, is the process of reaching out to the community and involving people in community access television. The type of outreach described here has two purposes:

1. Outreach should ensure that the center attracts a broad cross-section of people to be program producers and users. These people should come from a diverse community base, including all the different types of people in the community and all the varied interests of the people who are served by the channel.

2. Outreach should attract viewers, by encouraging people to watch the programs created and cablecast by community members with interests which are similar to their own.

We would like to underscore our strong opinion that community outreach is one of the crucial make-or-break devices that determine the success or failure of an access center.

Too often, an access center is understaffed, and thus does not have a well-organized outreach plan. What little community outreach does get done happens on a hit-or-miss basis, because there are so many people to train, so much equipment to keep running, and so many political fences to keep mended. All of these tasks, of course, are attempted by too few people with not enough hours to do them.

Outreach, then, keeps getting pushed further and further back on the list of things that must be done. All too many public access centers in this country are run using crisis management, as a result of lack of funding and, consequently, lack of staff.

The effectiveness of the outreach that *is* done, it must be emphatically stated, directly affects the livelihood of the access center. The types of programming, the user base, and the support base in the community are all critically influenced by outreach.

It is therefore vital for an access center to have some kind of an organized outreach plan firmly in place; the plan should have very specific goals and objectives. Such a plan should schedule the access manager and members of the access staff to address a certain number of community groups each month. It is also very important that the access manager does not do all of the outreach. This function belongs to and concerns all of the center staff—both paid and volunteer—every member of the board, and everyone involved in the center who meets the public in any way. This is true whether the meetings with community members are part of an outreach program, whether they occur during leisure time, or whether they concern community or family interests. Everyone involved in public access should be an advocate and should encourage others to become involved in access.

Individuals who are already interested in making television programs, for whatever reasons, need not be targets for outreach. They will beat their own path to the public access center.

Community groups, on the other hand, should be the primary focus of the center's outreach endeavors. Community groups have an identifiable constituency and a set of goals and objectives, so their involvement is critical to the success of the access operation. When a community group agrees to work in access, all the members of the organization are brought in—perhaps a hundred or more people. Bringing in many such groups—each with its own production team of maybe a dozen people—and providing them with access training builds a base of groups from within the geographical area that have their own reasons and agendas for using the access channels.

Focusing on community groups accomplishes one of the goals of access outreach by building a broad user base. At the same time, this type of participation develops a broad base of viewers, because, instead of one new viewer, the community group offers many new potential viewers. And as the new group uses the access facilities, the members will promote the program that they are making or have made by using the group's promotion mechanisms, newsletters, and meetings.

Suddenly, because one group has done a television production, the access channel has a multitude of new viewers. Initially, many first-time viewers may expect to watch only the group's program, but once they see community activity programming, such viewers will stay with the channel for other programs. In this way, outreach goal number two is achieved. Bringing more users and viewers to access will diversify the user and viewer base.

How is Outreach Done?

Outreach should take a two-pronged approach to the community. The initial objective of access outreach is to make the community more aware that public, educational, and municipal access channels exist on their cable television system. See Chapter 12 for some interesting viewing statistics. Simply because a home has cable television (which may have been subscribed to solely to receive HBO) does not guarantee that the occupants know about public access channels.

It should be made abundantly clear to prospective new viewers how widely access programming differs from the commercial television programming that they are used to viewing—or indeed, not viewing.

Another major objective of community outreach is to describe how the video medium can be used by community members, both as individuals and as members of community organizations, to implement their group or individual agendas.

Outreach accomplishes its aims by developing and maintaining contacts with individuals and groups in the community. To develop contacts, one must literally reach out into the community and come face to face with its leaders, doers, and movers and shakers.

Developing Contacts

One of the more practical ways of developing community contacts is to talk to representatives of community organizations. Every community has many such groups, for example:

- Human services organizations—the Red Cross, Planned Parenthood, community mental health agencies, and the American Association of Retired Persons.
- Service clubs—Kiwanis, Lions, Optimists, Odd Fellows.
- Special-interest clubs—stamp collectors, ham radio enthusiasts, computer user groups, and camera buffs.
- Religious organizations—churches, synagogues, and the local ministerial group.
- Youth organizations—Girl and Boy Scouts, and Campfire.

Virtually every local library has a listing of these groups. Privately printed local telephone books often list community groups, with their key members' numbers to call.

Call or write to these groups and arrange to attend one of their meetings as a speaker. Better still, invite every organization to send a representative to a meeting on community access. Ask for an RSVP so you'll know how many people are coming and will have enough chairs, coffee, and so on. At this meeting, describe public access and its philosophy, and field questions about public access cable television and what it can do for the community's organizations.

Explain that the organizations and their community efforts can get public exposure on the access channel. If at all possible, have sample access programs available and show portions of them. By all means have enough membership applications for your organization and printed information on how to become an access producer so that some can be taken back to the community organizations.

Also invite the organizations to join the access corporation and to send people to be trained and to do programming. Every community organization has a person who does public relations for the group. If you can reach this person and convince her that access will benefit her group, your job with that organization is complete. The public relations person will do the rest; you have provided the vehicle.

Community outreach is a never-ending job, and therefore one such meeting will not be enough. Organizations in the community change members and add new ones, and new organizations spring up daily.

The community outreach effort of the access center may also be used to gather audience ascertainment information and to provide a public relations tool in selling the center to the community.

Individual outreach must be considered as well, although on a lesser scale. As we have said, most program producers will naturally gravitate to the access center, but an effort must be made to enlighten and educate those in the community who are not aware of the center's existence. Outreach to the individual can be done by highlighting access production in the local print media; by doing shoots at local shopping centers, preferably during busy periods like Friday evening or Saturday afternoon; and by explaining access, perhaps through a well-written pamphlet, to all who ask. Individual outreach can also be performed by inviting new access viewers on the cable system to use access, and by having staff members promote access whenever possible.

PUBLIC RELATIONS AND PROMOTION

Community outreach and public relations (PR) efforts are two sides of the same coin, inasmuch as public relations can be defined as letting the community know who you are, what you are doing, and what your goals are. This same definition can be applied to outreach.

The difference between the two lies in the techniques that are used in public relations. These techniques are designed to increase the visibility of the access center and its programming within the community. It is very important to understand that—no matter how good the center's programming is—if virtually nobody watches it, then the programming is created in vain.

Public relations efforts include writing press releases about access programs for the print media, getting the releases printed in the local newspapers, training program producers to promote their own programming, and getting local community groups who put on programs to use their own promotional facilities.

Another PR technique widely used at access centers is the newsletter. These newsletters are usually published monthly or quarterly, and sometimes they contain advertising paid for by local merchants to defray costs. The newsletters contain information about the access center, columns by the board president and by the access manager, news about upcoming activities and productions, and invitations to the community to join in the fun. The newsletters are mailed to center members, and placed in public places such as libraries and supermarkets where people can pick them up. The newsletter can often benefit from the talents of the writers so frequently found in the access volunteer cadre.

A third technique is the access program guide, which is sometimes in the newsletter, and sometimes published separately on a monthly basis. The guide lists the access programs currently running on the channel, in much the same way as the daily television guide in the newspapers. Indeed, local weekly newspapers should be asked to print the guide as a public service.

To be successful, the guide should be broadly distributed within the community. In this context, a fourth technique would be the use of the cable company's monthly mailings. Distribution can be greatly aided by help from the cable operator, who can include the guide with the mailing of monthly cable system billings, or can include or incorporate it with his cable system program guide. This guide is either printed and distributed by the cable company or by a program guide company—a printing company that prints and distributes guides for a number of cable companies. In either case, the guide goes monthly to every cable subscriber. One such successful program guide that has been running for a long time is the guide produced by the cable company-operated access organization in Arlington, Massachusetts.

In order to attract viewers, we consider it critical to the access center to list its programming in the cable company's guide. It is admittedly difficult to list individual access programs, because cable company guides are printed six to eight weeks before they are published. Most access centers are not able to schedule programming so far in advance—although we consider it important to make the effort. However, program blocks can be created and given generic names instead of specific program names, such as *Local Sports, Community Churches,* or *Community Montage.* Every week the same time slot offers a different program with a different community group, but with a similar program theme. This technique allows the center to insert a name for the viewers to see in the cable company guide. The bottom line is that, in order to get the community to view access, the access programs must be listed in the guide. In addition, if the funding is available, it is a good idea to include a display advertisement in the guide—perhaps in a box each month that features an access program and behind the scenes information about it, or that describes a service of the access center, or that highlights some of the months' programming.

A fifth PR technique frequently used is to design colorful access center logos, to be printed on bumper stickers and buttons and used in promotional campaigns ("promos") for access activities.

A sixth method of public relations for the access channel is to use the "spot announcement" or ad availability time made available to the cable operator on many of the satellite services distributed by the cable company. With the consent of the cable company, this spot time can be used for public service announcements (PSAs) about the access center. Using this technique, tools of cable television promote access or its specific programs. It is advisable to have the center's regular series access producers create PSAs about their programs. The cable operator would then be asked to run the PSAs at specific times on ad availability time on a satellite service. Most cable operators are not using nearly as much of this ad availability time as is available to them, and should be amenable to running PSAs that promote channels on their system.

It is also advisable to attempt to tailor the PSA to the satellite service on which it is run. For example, if the access channel features a regular rock music program, it makes good sense to run its PSA on MTV; a local sports program, like little league hockey, should run its PSAs on ESPN, the entertainment and sports network. On the other hand, a wide variety of PSAs could be run on CNN (Cable News Network) or USA network. A local health or fitness program could be advertised on Lifetime.

All of these techniques can be used to constantly increase awareness of access in the community, and to foster a good public image of the access channel in the eyes of the community.

ACCESS ROLES AND RELATIONSHIPS

A number of key relationships must be fostered and maintained in order to ensure successful access. The roles that the access center staff—and particularly its manager—must play with respect to access users, cable subscribers, local government, and most certainly the cable company are of ultimate importance.

Relationships with Users

In their relations with access users, the access staff should always present a positive attitude to the uncertain beginner who knocks on the access center door. This initial support is followed by continuing words and gestures of encouragement to reassure the neophyte that a newcomer can learn the techniques, skills, and dexterity that are required in access television operations. The positive approach continues with up-beat constructive criticism of programs that could have been better and will be in the future.

There should always be an understanding that the access user is in a safe place, where his or her personal beliefs can be developed into program themes without scorn, derision, or nay-saying on the part of access staff members. Relationships with the access user are finalized by an honest invitation to the newcomer to become a member of the access center family.

Relationships with Cable Subscribers

The cable subscribers should be continually invited to watch programs on the access channel. They should be educated in the purposes of public access, and they should be enlisted to make their own programs. They should be sent access buttons and bumper stickers as part of public relations campaigns. Most important, they should be made aware of the fact that the access corporation does not censor programs. If, unfortunately, subscribers are exposed to a program that offends them, they should understand the implications under the First Amendment, and should not blame the messenger for the message.

Relationships with Local Government

The access corporation must take a broad and encompassing approach to its community's local government. The local government needs to realize that public access programming is important to the community. Government should support the concept of access because the role it plays exemplifies the essence of the First Amendment. In many instances in which local government has not been supportive, the cause has been a misunderstanding of access's role. Occasionally, local governments have felt threatened by access, or because access was not controlled by government, city fathers have become nervous and uncomfortable about access.

The access corporation should support the efforts of municipal government, particularly in the fields of law enforcement and fire protection, and it should cooperate with government, especially if there is no municipal channel on cable. Further, the corporation should promote a spirit of teamwork with the municipality and its citizens. It is sincerely hoped that the municipal government will respond in kind. Lines of communication between the access corporation and its government should always be diligently kept open. Access should always cooperate with city staff members—most particularly those who help enforce the provisions of the cable franchise agreement—and keep them abreast of anything that makes their jobs easier. Failure to concentrate on these relationships will assuredly result in the types of conflicts referred to in the next section (The Politics of Access) and in Chapter 11 (First Amendment Issues).

Relationships with the Cable Company

The most important relationship is between the access corporation and the cable company. These two entities have a great deal in common: the cable system, the subscribers, and the viewers. It should be noted and underscored that the more access users there are, the more subscribers to cable there will be—and, of course, the more dollars on the profit side of the cable company's ledger. Although it is necessary to be businesslike in relationships with the cable company, it is also necessary to be friendly and cooperative with its general manager, operations manager, and technical staff. These are a few of the people who can either make life very comfortable for the access manager or make the manager's job a living hell by dragging their feet on staffing, equipment servicing, and the many other ways in which the company and access interact.

THE POLITICS OF ACCESS

Playing the politics of access can be the most crucial aspect of outreach that the access manager can perform. Specifically stated, the politics of access concerns keeping the local community's politicians aware of and involved in public access. Further, their involvement must start at the very beginnings of the access center. If the center's outreach and public relations efforts reach everyone in the community except the local politicians, then access may very well be doomed in that community. Politicians are rarely neutral; they're either for you or against you.

Outreach to politicians must include describing the access philosophy, letting them know how access works, and telling them who does access in the community. The politicians should be individually invited, not only to appear on access programming, but to do access programs about their roles in the community. The outgoing nature of a politician makes him or her a perfect access user.

Next we describe an absolutely hypothetical example of what can happen in a community when the politics of access are not played positively. We present it in narrative form, and we repeat for clarity the disclaimer that our example is indeed fictitious.

In a certain small town somewhere in suburban America, there was only one medium that molded public opinion before cable television and public access arrived on the scene. That opinion-maker was the town newspaper. It was, perhaps, a typical small town paper, whose main virtues were its coverage of local events and its ads for local merchants.

This weekly paper tended to use editorials to villify everything and everyone in the town with which it, or its editor, did not agree. The citizens of the community, by and large, went along with the views of the paper, since they were not privy to any other publicly disseminated point of view.

The local politicians and would-be politicians in this town learned pragmatically and early in their political education, that it would be difficult for them to be elected to local public office if this paper editorialized against them. The politicians, therefore, tended to pander to the newspaper and to its editor and publisher when she took an editorial stand.

Then public access cable television entered the picture. Its access manager, who simply assumed that the town politicos were proaccess, asked the paper for support in publicizing access. The editor of the paper, perhaps jealous of her monopoly on public opinion, indicated that she was not interested. She wouldn't publish the access program schedules and she wouldn't list access meetings, although every other club meeting in town was noted in the paper.

Then, from out of the blue, while access was young, growing, and relatively helpless in a PR sense, the front page of the paper quoted a member of the town council who excoriated the access organization and its manager for not covering an event at the local community center. Of course, the access people were not asked to cover the event. And even if they had been, it was explained to the councilman, that's not how access works: public access television is not NBC or CBS. Regardless of the truth, the councilman's outburst was followed with an editorial in the next issue of the paper, agreeing with the councilman and decrying the ineptitude of the access organization and its manager.

A member of the access board wrote an explanatory letter to the editor, which was published in a later edition, and the furor seemingly died. Alas, not so; within a few weeks, the same councilman called on his council colleagues to rescind the city proclamation that had brought access into being and that, more importantly, provided the access organization with its two-thirds of the five percent franchise fee that the cable company paid to the city. This fee was the primary source of revenue available to support access.

Thus the editor and publisher in our hypothetical tale was presumably able to influence the councilman so that he would unobtrusively kill off her perceived opposition and preserve her monopoly on the minds of the citizens. And at the same time, the citizens' view of local government was not going to be affected by the light of television programming.

How does the story end? It could end in any way. The point is that tragic consequences must be prevented in the first instance by carefully putting and keeping the politicos on access's side. It should be easy, before the fact, to get them to understand that public access can be a valuable community asset, and that the people of the community will come down on them like a ton of bricks if any political actions that inhibit the free speech of the community are made known. Sometimes the access manager has to play hardball!

EDUCATIONAL AND MUNICIPAL ACCESS

The cable law of 1984 speaks of three types of access channels: P, E, and G, or PEG. The P denotes public access, E stands for educational access, and G describes government or municipal access. This book has so far concentrated on public access, but this chapter delves into educational and governmental access and examines how they differ from public access.

EDUCATIONAL ACCESS

At this juncture we will discuss the use that educators in the community make of the access channels at the preschool, elementary, secondary, college, higher education, and continuing education levels.

Essentially, three scenarios depict how access is used by educators; which scenario is in effect is mainly dictated by the size of the cable system and when the franchise was granted. In the first scenario, educators share a single community channel with public access. This is often the case on the older, smaller cable systems with less than thirty-five channels. In the second, the franchise was probably granted between 1973 and 1978, and so one or more channels may be specifically dedicated to educational use. In this case, the community educators have usually formed a consortium to operate the channel and decide how the time will be divided and programmed on the channel or channels. In the third scenario, the franchise will have been granted after 1978 or 1979, so that a large number of channels are available on the system. In this case, several educational access channels will often have been granted: one for kindergarten through high school educators and one for post-secondary educators, or—even more often—channels dedicated to specific institutions, such as a community college or a local university.

Because each of the scenarios developed at a different time and under different circumstances, the programming and management structures that were developed for each will vary.

Let us, then, talk about educational access channels in communities where they are separate from P and G. That is to say, the communities with specifically designated channels for education. For the most part, these are communities that have granted franchises since the mid-1970s, and in which the education channels are not designated for use by a particular school, school system, or community college. Rather, the channels were assigned based on the organizing abilities of the educator groups in the community, or often on the ability of the organizing educator who was first placed in charge of the channels and who was responsible for their development. In some instances, the elementary and secondary (kindergarten through high school) educators and the postsecondary educators form an education access committee composed of their media specialists, librarians, and audio-visual experts. This committee makes the decisions about which school uses which channel and at what times of day. Usually, elementary and secondary have greater usage requirements during the school day, and postsecondary use is needed more during the very early morning, the late afternoon, and the evening.

Sometimes a particular school, school district, or community college will be specifically designated in the franchise document as the holder of a particular educational access channel. As we have indicated, this scenario will occur most frequently in areas where franchises have been granted since 1978. Often in these cases, the public school system and the community college will each have been granted a channel, and the franchise agreement is the mechanism that determines who will receive the educational channels.

School systems or school boards use access in three specific ways:

1. for bringing information about the schools or the school systems to the community;

2. for "inservice training"—the education profession's term for updating teachers' skills, often called professional development; and

3. for instructing using the television medium.

Informational Use

The community can be informed about its schools through video programming—that is, programs transmitted live or on video tapes; or it can

FIGURE 9.1 Young access cameraman expressing pleasure as he shoulders a video camera for the first time. Courtesy of the NFLCP photo files.

be informed more simply, between programs, by using a character generator to place an easily updated school bulletin board on the educational access channel. The character generator's alphanumeric information is delivered quickly, continuously, and directly to an audience that may be waiting anxiously to learn if the weather has cancelled or delayed school on a wintry day or if the school baseball team has won its game with its arch-rival from the next town. The audience may also read school lunch menus, school event calendars, and notices of all kinds.

Informational video programming can run the entire gamut of school activities, both curricular and extracurricular. Plays, assemblies, club meetings, and school orchestra or band concerts might be videotaped. Video programming can certainly include sports events.

The programs can also feature interviews on current and pertinent topics with the school nurse, school administrators, and often with members of the board of education. Some schools do news programs about what is happening at the school, providing a creative form of expression for both students and faculty.

Students eagerly vie for positions on the studio crews that do the productions, as well as for on-camera positions (see Figure 9.1). Some schools, on the other hand, veer away from using students as either cast or crew, because of a mistaken belief that their programs must look as good as those seen on PBS. Clearly, student productions will never approach the quality of PBS or the commercial networks. But their less-than-perfect programming does not reflect negatively on the schools; instead, student productions have a spontaneity that transcends production quality. And making programs raises the electronic media literacy level of elementary and secondary students by teaching them to use the tools of television production.

Statistics show that the average American watches 7.8 hours of television a day. The average student, particularly below eighth grade, watches more than that. If the school systems can raise the television literacy of students, they will become better, more selective, and more intelligent television viewers. The end result might even affect the type and quality of programming offered by the television networks.

In educational programming—as in all other access programming—adequate concern should be given to the legal constraints of access. Completed talent release forms should be received from all on-camera participants before the cablecast of the program, particularly those people who have been videotaped outside of the studio. Before using any material that requires permission, its copyright clearance should be arranged; otherwise the material should not be used.

It is important to the community—particularly if there is only one educational channel—that secondary schools or community colleges do not excessively use the channel to the detriment of the elementary schools. This can easily happen if a college has a communications education program, or if a high school is permitted to crowd the channel with its sports activities.

Elementary school programs, such as spelling bees between similar grades at different community schools, with PTA parent groups supplying the production crews, are sure to arouse great community interest during viewing hours (see Figure 9.2). Homework help programs for elementary and secondary groups has become popular, and such programs are proliferating throughout the United States. In homework help programs, dedicated teachers or senior students on a television panel help tutor younger students in subjects such as mathematics, history, or English. In Beaverton, Oregon, a group of four high school seniors sits on a cable panel and answers homework questions called in by telephone from students of all grades. If the panel does not have an answer, then teachers may be consulted, or the answer may be given later in the program.

This type of program is not designed as a Trivial Pursuits type of game, nor is its purpose to stump the panel. Rather, the program is intended to

FIGURE 9.2 Videotaping an elementary school classroom activity. Courtesy of the NFLCP photo files.

offer genuine help with homework problems. The first such program began a number of years ago in Irvine, California, and it averaged 100 calls a day before it left the channel because of lack of funding.

Inservice Use

The teaching profession should use the educational channel to broaden the scope of inservice activities. The traditional inservice training for teachers usually consists of teachers listening to lecture material that runs on ad nauseum with minimal value—at least according to comments elicited by these authors. The same material can be delivered in a more interesting video format and can be shortened to accommodate the acknowledged human attention span of fifteen to thirty minutes without a break or a change of pace.

Further, the education system can build a library collection of tapes on topics that really concern teachers and deliver that taped material via the cable system to places like teacher's lounges at times when teachers are willing and available to view such information.

Locally produced tapes by experts on particular course subjects or on particular teaching skills are especially welcome. Local production also involves teachers in the rewarding activity of creating television. Videotapes so produced can be exchanged with neighboring schools to both build tape libraries and to share expertise.

School nurses might use the inservice facilities to build a relationship with the local hospitals(s) by having the hospitals produce inservice programs, perhaps on a monthly basis, to update nursing skills, describe procedures for handling flu epidemics, and warn nurses of symptoms that they should look for at their schools. Television can be invaluable for this type of inservice, because it permits the nurse to be at her school duty station at the same time that she is updating her skills. If her training is interrupted by a medical crisis, she can use the school's VTR to record the program so that she can resume her updating when the crisis is over.

Instructional Use

Instruction using the video medium has many facets. It has been around long enough at the national level that most people with a television have been exposed to some form of televised instruction as seen on the Public Broadcasting System. PBS refers to televised instructional programs as "telecredit courses," since they can be taken for college credits through mail enrollment at the colleges.

The typical student taking college-level telecredit courses is female, over twenty-five years of age, white, and is in continuing education. Telecredit courses are usually chosen by students without consideration of the distance between their homes and the campus, but with consideration of the time frame within which the telecourses are offered. Most telecredit students have full-time jobs and need courses that are available outside working hours. These times include very early in the morning and after the usual working day is completed.

The most popular telecredit courses at one community college in the Washington, DC, area are business, data processing, management, English writing, and US history—in that order.

Some universities or college consortiums are seriously considering encoding or scrambling educational access channels. This appears to be a contradiction in terms, because encoding an educational access channel would no longer make that channel accessible to every viewer on the system. From the educational institution's viewpoint, however, an open channel gives away instructional information for which matriculated students on campus are required to pay. These college fees support the institution and make its instruction viable. The alternatives under consideration are: to use the channel unencoded for parts of the day, and encoded on a pay-per-view basis for other parts of the day; to make some educational access channels completely encoded and leave others completely unencoded; or to encode leased channels instead of access channels to deliver education on a pay-per-view basis for telecourses that the universities deem too expensive to deliver without cost. The technology exists to allow the cable operator, in conjunction with the educational institution, to encode any channel on the cable system—like the premium channels—either on a pay-per-month or pay-per-view basis. What then needs to be worked out between the cable operator and the educational institution is the method for approving individual decoding and for transferring payments for the course from the cable operator to the educational institution.

Using cable television to deliver instruction gives the prospective student an even greater choice of available courses. Using tapes, local community school systems can make courses available on cable with much less expenditure than the equivalent instruction given in the traditional classroom setting. Courses such as high school fifth-year Latin, which has only a very few students at any one school, can be delivered to the students of all of the schools in a county at the same time, using only one Latin teacher.

At least fourteen universities or colleges in major population centers in the United States have used ITV (instructional television) systems since the early 1980s. These systems combine studio classrooms with three-camera facilities and transmit their educational programming in real time to off-campus and distant satellite classrooms. The off-campus classrooms, replete with large TV monitors and telephones for student interaction, receive the programming on the ITFS band (Instructional Television Fixed Services) at about 2500 to 2700 megaHertz. There are thirty-one such frequencies in the band, authorized by the FCC, each with a power limit that constricts their broadcast radius to about thirty miles. Where cable television has arrived in their communities, a few of these institutions are integrating the ITV system with the cable system's educational channels to expand the ability of the university to provide higher education to the community.

From the educational traditionalist's point of view, this system perpetuates—although by television transmission—the standard teacher-student classroom system. From a television production viewpoint, which is where this book stands, the system also perpetuates the talking head concept of programming, which all media practitioners agree is the dullest possible type of programming. Talking heads are as effective as a teaching tool as using audio alone.

Using television as an education delivery system permits the use of videotaped slides and films, which are traditional additions to the lecture, but it also permits the use of animation and original videotaped material to replace static printed graphs and charts. In fact, television allows all of the facilities of a dynamic moving medium to be used in the educational program. These facilities include the optical tricks that can be done with colorizing, mixing and wiping, and the alphanumerics of character generation. Such features are not merely gimmickry applied on top of solid educational material: they contribute to educational delivery by breaking up the boredom felt by students who have to stare at a lecturing instructor. Innovative video techniques make the material being taught come alive, with the result that the grades of the students and the quality of teaching become much higher. One would hope, therefore, that educators will begin to learn the scope of the television medium so that they can make greater use of these devices in the teaching and learning process.

A further suggestion to educators: Before attempting to produce a telecourse, scan the field for already-produced material that meets your needs. Telecourse production can be extremely expensive. If, on the other hand, a telecourse has to be produced, we suggest that the producer takes a more innovative approach to the program than has been taken in the past.

Educational access has been much slower in coming to flower than its public or municipal counterparts. This is apparently because the traditional educational bureaucracy must be convinced that it is all right to augment the traditional teacher-classroom methods using cable channels—or any other nontraditional method, for that matter.

We have cited here communities such as Irvine, California; Beaverton, Oregon; and Chicago, Illinois, which have dared to be innovative and have placed the needs of their students above their own tendencies to be traditional.

Faced with the misconception that television instruction would replace teachers, educational television had a hostile beginning more than thirty years ago. Teachers' unions, given the choice between funding higher teachers' salaries or spending for the tools of television, logically opted for their members' salaries. In the late 1960s and early 1970s, when substantial federal funding was made available for public education, school systems invested heavily in EIAJ (Electronic Industries Association of Japan) half-inch, black-and-white porta-pak television equipment. Sadly, that equipment now lies idle in school closets.

Many school administrators have poorly focused ideas about educational television, and limit its use to PBS programs in the classroom. Computer applications at public schools have advanced much faster than video applications, because computers are perceived as a mathematics tool and have the schools' math departments as a constituency. Video, on the other hand, is an interdisciplinary tool with no specific constituency; its use is spread across the spectrum of elementary teaching and the secondary departments that teach language, arts, science, math, vocational skills, and social studies.

Despite video's obvious benefits to education, the bottom line of a school budget clearly shows that chalk is cheaper than videotape. It is our sincere hope that community educators will rise above all of the drawbacks necessarily portrayed here and use educational access as the viable and valuable tool that it can be.

Appendix A at the end of this book presents a July 1985 Survey of Cable Television Utilization by Colleges and Universities.

MUNICIPAL ACCESS

We refer to government access as municipal access because that terminology better describes the level of government that employs access. Philosophically stated, municipal access can be managed from one of two perspectives: It can be used as a forum for distributing information about the activities of the local government to the public. It can also be operated essentially as a propaganda tool for municipal government, with no negative connotation ascribed here to the term propaganda. Most municipal access channels are operated more from a public information than a propaganda perspective. A municipal channel operated as a forum for the distribution of information to the public allows a diversity of viewpoints on public issues to be aired; whereas a propaganda channel only dispenses the opinion of the local government.

In a program about a typical local governmental issue (taxes, animal control, parking, etc.), it is safe to say that not all of the members of a city council will agree on that particular issue. All of the different opinions should be represented when the issue is discussed on the municipal channel. Ideally, the range of opinions among the community's citizens should also be reflected on the channel.

Other municipal access functions that differentiate it from public access are: informing the population about municipal government activities and concerns; making local government and all of its departments and services more accessible and understandable to its citizens (so states local government); and speeding up and improving internal government communication.

In some communities some municipal access functions are so similar to public access that it often becomes difficult to differentiate between the two. This situation usually occurs in communities where there is no public access programming, or where public access channels are underutilized for whatever reason.

Because municipal access is run by government, its rules of operation are different from public access. Of necessity, municipal access often operates under a more stringent set of internal rules than either educational or public access. For example, some regulations that might be included in a set of municipal access rules and procedures are:

- Priorities for production and cablecast time are not first come, first serve, but are prescribed by the municipality.
- Content restrictions limit programming to official municipal concerns.
- Policies carefully control and sometimes prohibit program editing (as with "gavel to gavel" coverage of public meetings).
- Technical standards are kept much higher than those normally allowed in public access, in order to qualify for government status.
- Rules governing the appearance—or sometimes, and more often the nonappearance—of candidates for public office.
- Rigid policies control the use of municipally owned production equipment.

Examples of Rules and Procedures for Government Access Channels

Here are the rules and procedures adopted by a couple of typical government access channels:

IOWA CITY, IOWA
GOVERNMENT ACCESS CHANNEL 29 GUIDELINES

1. The City itself will not undertake or sponsor any political programming on the Government Access Channel with the exception of providing information on any ballot issue if initiated and approved in advance by the City Council.

2. Political programming produced or presented by independent producers may be cablecast on the government access channel if it involves any of the following:
   A. A political candidates' forum or debate where all candidates for a particular public office have the opportunity to participate.
   B. A public forum on a ballot issue where all sides have the opportunity to participate.

3. Direct public access to the Government Access Channel for political programming by individual candidates, political party representatives, or supporters of any candidate issue (except as provided in #1 above) will not be provided. Time is available for such programs on the Public Access Channel 26.

4. The following list of general priorities will apply to the City's use of the Government Access Channel. These are recommended priorities to which adherence should be solely at the discretion of the Mayor and City Manager or his/her designee.

A. Programming of an emergency nature involving public health or safety matters.

B. Programming initiated by the City Council.

C. Programming initiated by the City Manager.

D. Programming from the various city departments.

E. Programming from city boards, commissions and other bodies of a similar nature.

F. Programming produced by a source other than the City which pertains to relevant local government concerns.

G. Other government related programming or pertinent non-local programs from regional, State, or Federal levels and/or sources.

5. Management and programming of the Government Access Channel will be under the authority of the City Manager or his/her designee. Decisions of the City Manager may be appealed to the Broadband Telecommunications Commission as provided in Ordinance 78-2917, Section 14-64.

6. Definition: For the purpose of this document, the definition of "political programming" shall be any programming which involves the endorsement of any political candidate, party, or ballot issue or participation by any political candidate or supporter(s) for the purpose of campaigning or otherwise soliciting public support for any candidate, party, or ballot issue in a political election.

*OPERATIONAL PROCEDURES FOR
BEVERLY HILLS MUNICIPAL GOVERNMENT
CABLE TELEVISION ACCESS CHANNEL 25/L*

A. Channel—The local government shall operate on Theta Cable Channel 25/L and shall use the designation City of Beverly Hills Government Channel 25/L Cable Television.

B. Modes of Cablecast—The City shall use five basic cablecasting modes:

1. Live cablecast—live coverage will be provided. Generally this will consist of cablecasts of City Council meetings and other selected public meetings and events of general community interest as designated by the City Council.

2. Tape delayed cablecast—many public meetings or events will be videotaped for cablecast at a later period. Some meetings such as City Council, will be cablecast both live and subsequently by tape at other convenient times during the week.

3. Locally produced programs—a number of programs will be produced locally to illustrate the functions or operations of some form of City government. These will include videotape tours of government facilities such as the library, recreation centers, parks, city facilities, or might be on specific City programs such as traffic improvement, streets, building inspection, or budget.

4. Outside resource programs—much material concerning local government operations is available elsewhere in the country and may be borrowed for local use. This is especially true of public safety programs and training films. This material will be used when appropriate to the City of Beverly Hills.

5. Datacaster Information Service—during all hours of operation when no other programming is scheduled, the datacaster information board will be used to provide a continuous display of current messages of interest to the public.

C. Access Policy—The local government channel is not the same as the public access channel which is provided by Theta Cable of California. Access to the local government channel shall be limited to City functions and operations. Any non-City request for access must be specifically authorized by the City Council.

1. Public meetings—all public meetings of City policy-making or advisory commissions or boards are authorized for cablecast. All taping or live cablecasting of such meetings must be approved by the City Council in advance. Formal City Council meetings will be taped and other meetings of various commissions and boards will be added to the schedule as time and manpower permit.

2. Informational programming—all City departments may submit requests for programming which they feel are appropriate for the government channel. These may be locally produced. Any programming requests shall be subject to review by the City Manager or his designated representative. Only those programs which are consistent with the overall operating policy of the cable channel shall be cablecast. City departments must verify in writing that they have obtained authorization to use any copyrighted material.

3. Individual statements—requests for access to the cable channel for the purpose of advocating a personal viewpoint or policy shall generally be denied unless part of an overall programming strategy to solicit personal interviews with equal time provisions for all. At some point a series of interviews or call-in

sessions with various elected or appointed officials may be scheduled as a regular part of the programming. Additional information and specific ground rules shall be established for any such programming prior to implementation. Announced candidates for City offices shall not be permitted to make personal statements over the City channel, except as may be part of formal public meetings, from the time of their announced candidacy until after the election, unless a scheduled series of statements from all candidates is programmed and equal time allowed to all. Any such programming would have prior ground rules and policy specifically established in advance.

4. Datacaster information board messages—information for the datacaster information board may be submitted by any city department. Messages submitted shall be commensurate with the intent of this policy statement. The City Manager or his designated representative shall be responsible for insuring the appropriateness of messages used.

5. Program log—a daily log will be kept to record all programming cablecast during that day.

D. Editing Policy

1. Public meetings—any public meeting cablecast shall not be edited or subjected to editorial comment. Meeting coverage shall be from gavel to gavel. Study session meetings such as The City Council Tuesday afternoon Study Session are not currently planned to be cablecast. Supplementary information on agenda items which will aid the viewer in understanding the issues may be provided if necessary.

2. Departmental programs—any programming prepared by or provided by an individual City department may be modified or edited as appropriate to the policies governing channel use, or as dictated by scheduling and manpower requirements. This process shall be under the operational direction of the Director of Community Services, with ultimate responsibility for program content resting with the City Manager.

3. Datacaster information board service—messages programmed into the datacaster information service shall be edited to provide clarity and to maximize use of the sixteen page memory bank currently available. The Director of Community Services shall have operational responsibility for this editing.

4. "The City of Beverly Hills, its officers, employees, and agents

shall not warrant the accuracy of any information cablecast over the cable channel."

E. Endorsements—at no time will the channel endorse specific brand names of products for consumer use.

F. Promotions—Promotional announcements for City-sponsored events are acceptable for cablecasting. Promotional announcements for events, charities or outside organizations in which the City has no official interest or sponsorship shall not be allowed.

G. Use of Outside Resources—In order to maximize programming, every attempt will be made to use outside personnel resources to assist in channel utilization. Specifically, an arrangement will be made with University of Southern California, University of California Los Angeles, Loyola Marymount University and Beverly Hills High School to provide intern students to perform cablecasting services for City Council meetings under the direction of the Director of Community Services and to assist at other times mutually agreed upon.

A mutually beneficial arrangement with Beverly Hills Unified School District will be made to allow for television programming materials to be placed on the City channel at various times. In return, this agreement would allow the School District use of government programs and tapes for use in educational programs. Tapes, programs or films from other sources which would be appropriate for local use will be sought and used where appropriate to supplement local programming.

H. Use of City Equipment—Use of City-owned video equipment shall be restricted to City activities and by City employees or under the direct supervision of City employees. Loaning of equipment for personal or outside use shall not be authorized.

I. Channel Operating Hours—It shall be the general goal of the City channel to have some form of programming available continuously. The general approach will be to utilize live and taped programming when available during weekdays, and to have a continuous running Datacaster Information Board Service at all other hours, 24 hours a day.

J. Retention of Tapes—It shall be a general policy to retain video tapes of locally produced events and meetings for a four week period. At the end of that time, the tapes will be reused and the original material erased. Any requests for longer retention should be made in advance of the four week limit to the Director of Community

Services and, if possible, a replacement tape made available. The tapes shall not be considered an official record of the meeting and there shall be no liability for inadvertent erasure or omissions.

Municipal Access Production Priorities

The production priorities of a municipal access channel might involve production decisions about requirements in the franchise agreement; they might limit the discretion of the municipal access staff; and they would include prioritized requests by other government agencies, such as the city or county council, the commissions and committees of the local government, the mayor, the city manger, the county executive, or by the cable regulatory commission.

As might be expected, in municipal access these priorities are at least somewhat political in nature. That is to say, a production request initiated by the mayor might receive a higher priority than one initiated by a city committee. On the other hand, an emergency program involving immediate public health or safety concerns would carry a higher priority than a mayoral request.

Content restrictions most often preclude "political programming," which can be defined as any programming that endorses a political candidate, a political party, or a ballot issue. This definition includes the participation of anyone who appears in a municipal access program for the purpose of campaigning or soliciting support for any candidate, any political party, or any ballot issue in a political election. Time factors, such as filing dates for political office, are often a part of the restrictions.

This is a major point of departure from public access, which usually endeavors to give equal time to all sides of an issue and tries to include all of the candidates for public office—or, at least, all candidates are invited to participate if one candidate for office is included in a program.

Municipal access regulations generally include a "no editing" policy when city or county council meetings are cablecast. During the cablecasting of municipal meetings, the cameras are usually required to stay on the municipal participant(s), as recognized by the moderator of the meeting, and are not to swing around to get audience reactions.

These restrictions do not deter the municipal access staff members, however, who accept them as a way of life. Often, staff on the municipal payroll and volunteer members from other departments of local government are able to produce sterling examples of access programming. This programming often includes call-in shows that feature local government figures answering citizens' questions, complaints, and concerns in real time. Programs that explore communities and measure the pulse of the various groups within a city have been expertly done on municipal access.

These programs may include oral histories of vanishing ethnic groups, mini-documentaries about classic city landmarks, stories of inner-city "sweat equity" home ownership (which enables tenant groups to purchase and manage their own buildings), stories of local heroes who miraculously rescue their fellows from home fires, and similar programs.

The municipal channel in Kansas City, Missouri, is operated by the city's Public Information Office. In a descriptive paper issued in November, 1982, their staff was listed as: The PIO director, two journalist 1s, a producer/engineer, a producer/director, a half-time graphics designer, a secretary, a clerk, and various volunteers who were mostly students. The following excerpt is quoted selectively from what they called "Some Things We Learned":

1. If your video equipment will get regular use, particularly in the field, you can expect it to need regular maintenance and to break down periodically, even if it's new. Not all malfunctions will be major, but if you care about the quality of the video and audio that you are producing, you will find the equipment in the shop more often than you think. We budget $1000 to $1500 a year for maintenance and have sometimes exceeded those figures.

2. Never underestimate the time it takes to shoot and edit videotape. We sometimes spend weeks or even months of on-again, off-again editing to finish a show, although the amount of effort that we put into programs may not be typical.

3. Try to get permission to use video equipment under your own working conditions before you buy it. One reason that we traded in our original camera, at a great loss, was that it wouldn't genlock, it could not be used in multiple camera taping, a fact which was overlooked when we bought the camera.

4. If you are preparing to seek an initial allocation from your city council for video equipment, avoid a bare bones approach. Without going overboard, take your best shot in your initial request, since you may not get much more equipment until your channel proves itself.

5. If your channel is very active, it will gobble up videotape. We bought 10 cases when we started. Within a year, to the astonishment of the budget office, we had to buy more. When a program is in progress, it can frequently tie up 10 to 25 cassettes temporarily for raw footage that will be used in the finished show. Each finished program also ties up a cassette, and you will probably want to create a file of footage for future use that will occupy still more tape.

6. Don't expect [other] city departments to beat a path to your door requesting TV shows. If it happens, good for you. What is more

likely, however, is that after an initial burst of interest by departments, you will have to generate a lot of the shows on your own. Access video doesn't have the glamour of commercial TV, and some of your more shortsighted administrators may think it's a downright waste of time. Prepare for a long-term sell job.

7. At the earliest possible point in your channel's development, begin building a network of outside [technical] resources you can turn to. Nothing is more scary than to schedule a TV show, then have your equipment break down the day before airtime.

8. Some of our best narrators, moderators and camera operators have come from the ranks of city employees. You will probably find at least a few who have had some training or experience in broadcasting, or at least a natural talent for it. We use these people as fully as we can. They add new faces to the channel and help relieve some of our workload. We also use volunteer professionals at times, but coordinating our schedule with theirs can greatly slow down production.

9. As your channel grows more active, you may find that you are expecting more staff and equipment support from your cable company than the company is willing to provide. We found the company wanting to back away from some of its earlier commitments. The cold fact is that access channels don't produce any direct revenue for the cable company and, in a crunch, access goes to the bottom of the company's priority list. So, be prepared to stand up for your rights, and to receive gentle pressure from the cable company to become more and more self-sufficient.

Although municipal access in Kansas City and some other cities is produced mainly by city staff operating out of a PIO (public information office), other cities provide access training, through an access coordinator on city staff, for specific departments—most notably, police and fire departments. These departments are then made responsible for public safety programming since part of their purpose is to educate the public about hazards that may be life-threatening. Fire departments would always rather raise public awareness of fire prevention than fight fires, and an access program is certainly a more dynamic way to accomplish this than something like a fire prevention poster contest.

Crime prevention is addressed by members of the police department, who produce programs concerning window latches, door locks, removing keys from parked cars, walking at night in unlit areas, and the many other methods that are typically used to prevent crime.

Having trained the department personnel to produce public safety or other municipal programming, the city's access coordinator is then responsible for cablecasting the programs on the channel.

Many cities have programs to provide civic volunteers to plant flowers, help at senior centers operated by the city, and so on. These volunteers are often used as crew members on municipal access programs.

Other Municipal Programming

In addition to the public safety programming produced by the police and fire departments, other municipal departments, such as parks and recreation, planning, and the office of the mayor or city manager produce municipal access programming. At budget time the city's finance department will almost certainly want to produce a series of programs about the city budgeting process, perhaps leading up to a live cablecast of the budget hearings that are normally held by the city council.

Location of Programming

This section might be termed: Which local government department produces municipal programming? Studies that have addressed this question indicate that there are as many variations as there are cities with municipal access channels. We cite a survey developed and prepared in the spring of 1984 for the National Federation of Local Cable Programmers (NFLCP) by Andy Beecher, the Program Director of the Metropolitan Area Communications Commission of Beaverton, Oregon. Thirty-five surveys were dispatched in February 1984 to government cable programmers, most of whom were represented at the 1983 National Convention of NFLCP. The surveys included questions on organizational structure, staffing, budgets, programming, policies, and scope of activities. Appendix 9A at the end of this chapter details the results of the survey.

Availability of Equipment

Departments in most city governments own a large amount of video equipment that was purchased for in-house use long before cable television arrived upon the scene. A manager starting up a municipal channel would do well to initiate a thorough survey of what video equipment the city owns and where it is located in terms of what department and what building. What are the specific make, model, and format of each piece of equipment, what is its condition, and how frequently is it used? Gathering this type of information gives the manager a list of equipment that might be used for municipal access and thus might help prevent overloading of the access equipment budget.

The next step—and this, it must be realized, is a touchy issue—is to determine the best approach to managing that equipment. Is a centralized equipment pool an option? Should the equipment continue to be assigned to particular departments? City departments, particularly in large cities, tend to be very possessive of what they perceive to be their property. They may not wish to part with that property, even if it is not being used or if it is underutilized.

A possible solution to this dilemma for the municipal access manger is to develop a set of policies for the loan of equipment from one city department to another. The key factor here is to sell the idea that, through borrowing or exchanging equipment, each department will have more equipment available to it than before. A centralized system for coordinating equipment use might be sold to the city departments on the grounds that more equipment, and potentially more sophisticated equipment, would become available to them. Also, if department personnel are trained by the access channel, their department's equipment would be used in a more professional manner.

Cities with Municipal Access

Some of the cities in the United States with successful municipal access programming are: Lakewood and Beverly Hills, California; Rochester, New York; Dallas, Texas; Madison, Wisconsin; Phoenix, Arizona; Kansas City, Missouri; Iowa City and Dubuque, Iowa; Tacoma and Yakima, Washington; and Portland and Beaverton, Oregon.

APPENDIX 9A

Spring, 1984

This survey, prepared for the National Federation of Local Cable Programmers, is intended to develop a better understanding of how local governments are utilizing cable television to inform their populations, to make government more accessible and understandable, and to streamline and otherwise improve internal communications. It is hoped that the survey, along with others, will encourage a dialogue among those who operate governmental cable communications agencies.

Thirty-five surveys were sent out during February of 1984 to governmental cable programming operators, primarily ones which were represented at the 1983 National Convention of the N.F.L.C.P. The surveys included questions on organizational structure, staffing, budget, programming, policies, and scope of activities. Since government programming was only in the planning stages in Sacramento and Mountain View, California at the time the surveys were completed, the responses from these communities were incomplete. Their limited responses are, however, incorporated into this survey, as are their addresses.
The survey was developed and prepared by Andy Beecher, of the Metropolitan Area Communications Commission. For further information, or if you would like to submit articles for publication in a newsletter for local government programmers, please contact: Andy Beecher, Programming Director, Metropolitan Area Communications Commission, 12655 S.W. Center St., Suite 390, Beaverton, OR 97005; or call (503)641-0218.

Note: We have seen one other recent survey on government access, completed in January of 1984 in Phoenix by that city's Office of Cable Communications. This excellent survey elicited 96 responses from 180 questionnaires sent out. Persons interested in those results should contact:

> Joetta Downs, Office of Cable Communications
> City of Phoenix, 251 W. Washington, Phoenix, AZ 85003

If anyone knows of other surveys of governmental communications agencies, please contact us.

Reproduced courtesy of the NFLCP files.

LOCAL GOVERNMENT CABLE TELEVISION PROGRAMMING SURVEY

Organizational Structure / Activities

Operated By...

Of the 18 agencies surveyed, the following represents the level of government or organization under which they operate:	City Government	12
	County Government	1
	City or City-County Consortium	3
	Non-Profit Organization	1
	Cable Operator	1

Of those 13 agencies operated by a city or county govern- ment, the following represents the department in which they function:	Mayor or City Manager	4
	Community Affairs/Relations	2
	Other (Human Resources, City Clerk, Communications, Public Works, Fire Bureau, General Services, Community and Economic Development)	7

Activities

Agencies engaged in both cable regulation and government programming: 12

Agencies engaged in cable regulation, and government, educational, and public access cable programming: 4

Channel Operation

Agency operates one government access cable channel: 12

Agency operates two government access cable channels: 1

Agency operates one local origination channel (cable operator) 1

Agency operates eleven channels (City of Portland has its own cable system) 1

Agency produces programming, but does not operate channel (New York's Channel L Working Group) 1

Staffing

	Full-time Paid Staff	Part-time Paid Staff	Part-time Paid Interns	Un-paid Interns	Volunteers
Arlington County	1	-	-	1	65
Iowa City	1	1	1	4	-
Irving	11	-	6	-	-
Madison	3	1	9	-	-
Miami Valley	6	-	2	-	-
New York City	1	-	-	15	-
Norman	2	3	-	-	-
Pittsburgh	11	1	-	-	-
Portland Fire Bur.	3	-	-	-	-
Southfield	3	4	-	-	-
Spokane	2	-	4	-	-
Tacoma	3	-	-	-	-
Washington Co. (MACC)	2	-	-	2	Many
Yakima	3	-	-	1	Many

Training Activities

	Yes	No
Agencies training government employees in TV production:	6	11
Agencies allowing & encouraging these employees to crew and/or produce programs:	8	9

LOCAL GOVERNMENT CABLE TELEVISION PROGRAMMING SURVEY

Agency-trained government employees participated in productions as crew and/or producers with the following frequency (of total agency productions):

Less than 25%	Approx. 25%	Approx. 50%	Approx. 75%	90 - 100%
2	2	1	1	2

The survey asked how these agencies decide what to produce:

Required by Franchise Agreement	3
Staff Discretion	15
Requests by Other Government Agencies	11
Requests by City Council / County Board	9
Requests by Commissions and Committees	7
Requests by Mayor or City Manager	11
Requests by Cable Regulatory Commission	2

Agency Budgets

	Provides Only Government Programming	Provides Cable Regulation and Government Programming	Provides Cable Regulation, and Both Government and Public Access Programming
$5,000 - 20,000	1	1	-
$20,000 - 35,000	-	1	-
$35,000 - 50,000	-	-	-
$50,000 - 75,000	1	2*	-
$75,000 - 100,000	-	1	1
$100,000 - 200,000	3	1	-
$250,000 +	1**	1	2

(* - One of these includes only those funds allocated for government programming;
** - Government programming is only a portion of this figure, which represents
an entire local origination department of a cable operator.)

Programming Activities

Public Meeting Coverage	3-4 /wk	2 /wk	1 /wk	2 /mo	1 /mo	Occa-sional	Planned
City Council	1	2	3	5	1	1	1
County Board	-	-	-	3	-	1	1
Board of Education	-	-	-	2	1	1	-
Cable Commission	-	-	-	-	3	1	-
Other Committees	1	-	-	3	2	3	-

Other Programming

	Occasional	Frequent
Production of Public Service Announcements	5	7
Programs Detailing Gov't. Services/Programs	4	6
Programs Identifying Employment Opportunities	3	1

Explanatory or "How-to" Programs; (utilizing the expertise of City or County agencies to develop community awareness):

LOCAL GOVERNMENT CABLE TELEVISION PROGRAMMING SURVEY

Department	Occasional	1x Per Month	2x Per Month	3-5 Per Month	6 or More x Per Month
Fire	8	2	-	-	3
Police	8	1	-	1	1
Library	4	-	-	3	-
Planning	5	2	2	1	1
Transportation	5	1	2	-	-
Health	3	-	1	2	-
Administration	5	1	1	2	1
City Council (Special)	2	1	-	2	2
Other	3	-	2	1	1
Live Call-in Programs on Public Issues	8	-	2	-	2
Public Issues Programs: No Call-in	4	1	-	1	-

	Yes	No
Elections: Candidates Forums & Interviews	10	8
Elections: Election Night Live Coverage	8	10

Programming Produced Within the Past Three Years

	Anchorage	Arl'ton Co	Iowa City	Irving	Madison	Miami Val.	Mtn. View	New York	Norman	Pittsburgh	Portl'd 1*	Portl'd 2*	Sacramento	Southfield	Spokane	Tacoma	MACC, OR	Yakima
Explan. of City Budget		Y	p	Y	Y	Y		Y							Y	Y		
Budget Hearings Coverage		Y	p	Y	Y	Y		Y	Y						Y	Y		
Explan. Codes/Ordinances		Y	Y	Y	Y	Y		Y		Y			Y	Y	Y			
Consumer Info.		Y	Y	Y		Y		Y	Y			Y		Y		Y		
Candidates Debates/Forums	Y	Y	Y	Y	Y	Y			Y			Y	Y		Y			
Fire Safety	Y		Y	Y	Y	Y	Y	Y	Y		Y		Y	Y	Y		Y	
Home Security			Y		Y		Y		Y				Y	Y	Y			
Consumer Issues			Y	Y		Y		Y	Y							Y		
Planning Issues	Y		Y	Y	Y	Y		Y	Y	Y	Y		Y	Y	Y	Y		
Health Issues		Y	Y	Y	Y	Y		Y					Y	Y	Y	Y		
Hearing Impaired			Y	Y		Y		Y										
Elderly		Y	Y	Y	Y	Y		Y		Y			Y	Y	Y			
Youth			Y		Y	Y	Y	Y		Y			Y			Y		
Physically Disabled		Y	Y	Y		Y	Y							Y	Y			
Minority Groups		Y	Y		Y		Y						Y	Y				

("Y" = Yes; "p" = Planned)

(* Portland:
1 - Residential Ch. by Rogers Cablesystems;
2 - Portland Fire Bureau, on City's own cable system)

Alphanumeric Programming (Character Generation)
The survey asked these agencies how effective character generation has been as a tool in government cable programming:

Extremely: 12;   Notably: 3;   Fairly: 1;   Poorly: 0

In-House Telecommunications
The following indicates the extent to which these agencies utilize institutional networks and/or residential cable systems for in-house programming purposes:

LOCAL GOVERNMENT CABLE TELEVISION PROGRAMMING SURVEY

	Yes	No	Developing
Use cable TV/video for in-house communications	12	4	-
- for in-service training of public employees	11	5	-
- for departmental communications; (briefings, teleconferences, etc.)	2	14	-
- to communicate to public employees new policies, benefits, etc.	2	14	-
- for employee orientations to jobs, etc.	6	10	1

- (Other uses listed include: transmitting city council meetings to department directors; producing departmental presentations for city council and staff.)

The survey asked these agencies to identify programs which they feel are unique, or particularly successful. Their comments follow:

Anchorage
"We have a tape on how to apply for a building permit which plays at the building permit counter. '82 and '83 have been record years for building in Anchorage, and the tape has apparently helped inform while people wait in line."

Arlington County, VA
"1) Periodic magazine-format program; ½ hour, 4 or 5 segments highlighting different county government services/activities. 2) County Board meetings - live."

Iowa City, IA
"Senior Citizens Health Series;...Employee Orientation Tape; Housing programs series (inspection laws for tenants and landlords); telephone breakup and how it affects citizens; Video Teaching series (production, philosophy, history of access); gardening series with County Extension Office; changes in tax laws; estate planning; changes in social services programs; buying/selling your home; Project Hard Times (Unemployment Counseling)."

Irving, Texas
"ICTN currently has 23 regularly scheduled programs plus monthly specials (i.e. performance of the Irving Symphony)...'Irving Game of the Week'--a weekly sports show highlighting local high school, college, and community sports from football to bowling to gymnastics. Over 11 different sports have been covered;..'Irving: Week in Review'--a weekly round-up of the news stories and topical issues which concern the citizens of Irving; 'On the Job'-- A local perspective on unique occupations. Topics have ranged from a magician to a mortician; 'Looking Up'-- a monthly show done in conjunction with the Irving Independent School District. A high school astro-instructor discusses the constellations and any unusual occurences of each month's sky; 'City Council/School Board-- cablecast live each meeting with rebroadcasting at later date."

Madison, Wisconsin
"CitiCable has recently begun to produce public service announcements for the various city departments. These spots are very much in demand and provide a cost savings factor to the departments; our services are nominal compared to a commercial production house. In addition, we are able to pass these p.s.a.'s on to the commercial stations for air time."

LOCAL GOVERNMENT CABLE TELEVISION PROGRAMMING SURVEY

Miami Valley, Ohio
"'Kettering Life' - the City's Parks & Recreation Dep't. has a volunteer who
produces reports on monthly activities; 'Moraine's Recreation Update' - same as
above, but produced by staff; 'La Dolce Vita' - sponsored by the Rose E. Miller
Senior Citizens Center, this live show concentrates on issues concerning seniors;
'Centerville Zoning Task Force' - a monthly program outlining zoning changes in
this rapidly changing city of 20,000."

New York City
"Channel L Working Group, Inc. was the first of its kind to initiate the live
phone-in programming with community boards, councilmembers and government
agencies. We incorporated in 1979, but have been producing since May, 1977.
Our format of one moderator and up to four guests has been lively, informative
and occasionally controversial... Each community board and councilmember rotate
on a weekly basis throughout the year. They aptly choose their own issues,
and guests have the opportunity to talk with viewers. Because of the diversity
of this city, topics covered are extensive and varied. Rent control, minority
job hunting, housing, ... rezoning... are popular topics."

Norman, Oklahoma
"We produce 'Working for You' which is a 5-minute program, cablecast weekly before
the City Council meeting and hosted by the City Manager. The show features
capital improvements around the city paid for by tax dollars. We have also
produced an employee orientation tape to familiarize new employees with policies,
procedures, and benefits... During flooding in October, 1983, we produced a show
on Emergency Management, including footage of the flood and the City's Emergency
Operation Center in action. 'Call City Hall', our only 'live' call in show to
date, was done before the City Council was to consider an ordinance concerning
discrimination because of sexual preference. The response was so great that we
extended the show from 60 minutes to 90 minutes. We intend to use this format
again in the near future on another controversial issue. We occasionally produce video for City Council meetings concerning items on the agenda. These are
mostly visual tours for the benefit of the Council that deal mainly with flood
damage, private roads, sewer lines, etc. We will soon be producing a 'Crime of
the Month' series with the Police Department and the high school. This program
will include a re-enactment of the crime using high school students as talent
and crew."

Pittsburgh, Pennsylvania

DEPARTMENT	TOPIC
1. Clean Cities Committee	Clean Cities anti-litter promo
2. Environmental Services	Animal Control and Refuse Pickup
3. Parks & Recreation	"Three Rivers Regatta", "Great Race", "Pittsburgh Zoo" and "Disabled Athletes"
4. Department of Finance	Informative tax program; city-owned properties for sale
5. Police Department	Neighborhood Blockwatch program / Project I.D.
6. Fire Department	Smoke alarms and kerosene heater safety
7. Urban Redevelopment Authority	Urban Redevelopment projects and properties for sale
8. Mayor's Office	Women's Task Force; Life Safety Alert System
9. Planning Commission	Mayor's 6-Year Development Plan
10. Cable Communications	Community Communications informational pgm.
11. Personnel	Job Listings
12. City Council	Live and taped coverage of legislative, committee and budget hearings.

<u>LOCAL GOVERNMENT CABLE TELEVISION PROGRAMMING SURVEY</u>

<u>Portland, Oregon (#1: Rogers Cablesystems)</u>
"City Council coverage has been contracted out to an independent producer;
cablecast citywide through an interconnect with another cable system."

<u>Portland, Oregon (#2: Portland Fire Bureau; (on City's cable system)</u>
"One program that is very successful is a live call-in show produced monthly
called 'Chief's Corner'. Firefighters can call and talk directly to the Chief
of the Bureau, who can dispel rumors and answer any questions."

<u>Southfield, Michigan</u>
"'Neighborhood Focus' - issues for neighborhoods: crime, taxes, Block Grants, etc.;
'Kids Choice' - library book reviews; musical events; yearly ice show"

<u>Spokane, Washington</u>
"'Safe City' - an interview format public safety program, featuring police and
fire officials; 'Spokane Insight' - a magazine format, film-style program
featuring timely segments on municipal events and issues; 'Art Beat' - on-location
cultural calendar featuring behind-the-scenes interviews; 'Energy Alternatives' -
film-style energy conservation program; 'Future Spokane' - roundtable discussions
about the big issues confronting Spokane's future; Election Coverage - the most
complete <u>local</u> coverage in town."

<u>Tacoma, Washington</u>
"City Council Meetings, PSA's. PSA's are produced for air on all broadcast, PBS
and cable stations in market covering three states and British Columbia;
'Crimestoppers' has been highly successful for Police Department and Crime
Prevention."

<u>Washington County, Oregon (M.A.C.C.)</u>
"'Washington County Public Affairs Forum' - a weekly program on local, regional,
state, national, and international issues, produced by M.A.C.C., employees of
local city and county agencies, and community volunteers; 'How Does Your Garden
Grow?' - a gardening informational program taped on location, in cooperation
with the Oregon State University Extension Service; 'County Line' - M.A.C.C.'s
first live call-in program featured county officials answering citizens' questions
on a variety of issues faced by our county government."

<u>Programming Evaluation</u>

The survey asked how government cable programming managers evaluate the effectiveness of individual programs or series. The responses were as follows:

Surveys sent to subscribers	3
(as part of a city newsletter)	1
Viewer calls and letters	5
"Live" phone-in programs, by number of calls	3
Addressable cable system can electronically survey the	1
number of viewers at any given time	
City/County staff comments to agency	2

<u>Advisory Committees</u>

Only five of the respondents indicated that they report to a government access
programming advisory committee. Of these committees, three are appointed by

p. 7

LOCAL GOVERNMENT CABLE TELEVISION PROGRAMMING SURVEY

city councils, one by a Mayor, and one by a cable regulatory commission (in the case of a consortium).

Government Access Policies

Nine agencies surveyed indicated that they have written policies for government access. These are: Iowa City, Irving, Madison, New York City, Norman, Pittsburgh, Southfield, Spokane, and Tacoma. Another set of rules was being developed in Washington County, Oregon. These rules include the following:

Policy Criteria	Number of Positive Responses
Priorities for Production	8
Priorities for Cablecast Times	3
Content Restrictions	8
Editing Policies	4
Technical Standards	6
Appearance of Candidates for Public Office	4
Program Sponsorship	1
Production Equipment Use Policies	2

ORIGINATING ACCESS PROGRAMMING

Access programming is intended to respond to specific community or individual needs. And those needs must either be formally ascertained through outreach or community studies, or they must be so clear and obvious that they cry out for a form of community expression. In either case, the access program producer who eventually addresses the needs will face a frustrating question: "I know what the community problem is, but how does one express the problem and perhaps its solution through the television medium?" In other words, he asks how community needs can be translated into a television program.

This book has already dealt with outreach and other survey methods, so this chapter is concerned with making programs that relate to serious, obvious community problems.

DETERMINING COMMUNITY PROGRAM NEEDS

An example of a community problem might be a local government that is unresponsive to the citizens' complaints about the wasteful use of taxes or about noxious smells from wastewater treatment centers. This type of local problem can be magnificently addressed by series of programs on an access channel that permit citizens to air their views for all of the community's response on a weekly or monthly basis. Concerns can be addressed, perhaps, by a telephone call-in program that includes real-time rebuttals or explanations by members of local government. A program of this nature fulfills the community's need to express itself to its governors in the interim between elections.

If the community or its members show little inclination to express community needs—as sometimes happens—then expressing them becomes the first job of a successful access manager. Boldly make your own assessments based upon your knowledge of the community or, alternatively, based upon the views of community members whom you can ask and trust. A newcomer to a community can also learn a good deal about its concerns by reading the letters to the editor of the local newspaper.

Having assessed what appear to be the community's needs, find one or more producers, assign crew members from among your trainees, and literally push them into doing a first production. Amazingly, you will find that momentum will build and other similar productions will follow. The manager will also find that if current programming does not live up to its promise to fulfill community needs, then some enterprising and courageous souls will go into the community with porta-pak equipment to ask "man on the street" questions—perhaps about abortion, forced school busing, or other pressing issues—and to make a program from the material they have videotaped.

HOW TO MAKE A PROGRAM

Once you have found a producer, he or she may ask how to make a program (as opposed to how to be a producer, which is a completely different subject). Here is a logical and systematic way to tell that person to begin:

- First determine the concern or issue. Research all aspects of it in depth. Then decide what you want the program's approach to the topic to be. This approach, or program concept, is created by identifying a prospective target audience and by picking a program format that should reach that audience. Is the concern best served by a panel discussion, a debate, a telethon, or a documentary? Start this process with the understanding that there is never a "sure thing" that is right for everyone or every subject. If one method of approach doesn't work, another can be tried until success is achieved.

- A critical step is researching the content, or the "meat" of the program. Know or learn everything about the topic. List on paper the objectives that the program hopes to achieve. Conduct brainstorming sessions with everyone connected with the program, and with everyone connected with the concern.

- Storyboard the proposed program if you use the storyboard approach to preproduction, and then prepare a program script followed by a shooting script.

- Start the preprogram activities with a logistic countdown. Arrange for the talent. Do a prelocation check if the show will be shot outside of the studio. Arrange for all production times, for checkout of location equipment, for studio facilities time, and for talent rehearsal.

- Work with the talent if a script is involved. Rehearse the program. If the show is to be conducted ad lib, do a "dry run" to try to anticipate any problems—either programmatic, technical, or talent-connected—that might occur during production.

- Tape the production. If the taping is done on location, view it there; if retakes are needed, they can then be done immediately under the same lighting and other scenic conditions.

- Carefully log all tapes as they are recorded and prepare an editing script. Then edit the tapes into a completed production, schedule the program, and cablecast it.

TELEVISION IN AMERICAN SOCIETY

Having described how to create a television program, let us now talk about another facet of television: where it fits into our society and what the access audience will—or should—expect from access programs.

Television is without a doubt the most compelling information force in society today. It has far outstripped the printed word in its communicative ability, its timeliness, and its use. Americans receive most of their news from television, sometimes merely moments after it happens, and they rely on newspapers and news magazines only for the in-depth aspects of stories.

For most of the readers of this book, television has been around all of their lives; it just is, and has always been there. Commercial and PBS broadcast television have not only always been there, they have also been virtually cost-free to the viewer with a television receiver. Because television forms part of our everyday background, the viewing public has developed a set of expectations about what television should look like. These expectations are based on what they have received and viewed *up until the emergence of cable access programming*. However, compared with the wide diversity in the packaging of print media, with its many formats and varied contents, the options presented to the television viewer have been very narrow and limited.

In part, the lack of diversity in traditional television programming is what the cable industry has used as its major marketing strategy. The cable industry has promoted the concept of specialized channels for movies, sports, news, weather, and children's programs.

When cable television arrives in an American home, viewing patterns begin to change as a result of this diversity of program format and content. Several studies have documented this fact.

Access and local cable programming provide the viewer with additional diversity in program content and format by focusing on local issues and local information. These services provide an opportunity for new, innovative, and often unique approaches to packaging program content. Such approaches provide television with diversity that more closely parallels the diversity we have come to associate with the print world.

Comparing access television to traditional broadcast television programming is like comparing apples and oranges—or, as some have said, apples and meat. Each has its own unique qualities that should not be slighted for the sake of comparison.

To be sure, access production entails compromises, such as small studios, a reduced number of cameras on a studio show, bare bones sets, and unpaid crew. Anyone considering these factors must realize the amount of expertise that is required in an access production, just to know how far one can compromise without jeopardizing efficient delivery of the program content.

Fortunately for those who must make comparisons, a number of studies offer documentary support of the belief that many television viewers are genuinely willing to accept access as a true alternative to traditional television.

Another given that can be positively ascribed to the television generation is that they have developed a strong program sense that has been achieved subliminally, so that people are unaware of its existence within them. This is another factor that must be considered by the access manager.

Although program content is often dreadful, the American television audience has developed, through constant and heavy viewing (seven to eight hours a day per person on average), an appreciation for the truly magnificent production of a commercial broadcast. The skills of all of the production personnel in the industry have been honed to a very sharp edge—the high salaries paid to such people demand it.

One very rarely sees a goof or a glitch, even on a live network program. Thus television audiences' technical expectations are so high that off-mike audio, snowy pictures, sloppy, jump-cut editing, and inadequate studio lighting are not acceptable on more than an occasional basis.

The message, we hasten to reiterate, is more important than the medium in access, but, at this time, a minimal technical level must be upheld if audiences are to continue to watch a television station. The level of audience sophistication constrains even the simplest of access programs to maintain a minimum cost of production so that programming doesn't fall below and stay below audience expectations. The audience *will* kill the messenger if they do not like the manner in which the message is delivered. We believe that this quality expectation has occurred simply because television's tools have never—until the advent of access—been available to the public. Providing the tools, providing a change, providing options are precisely what access cable television is all about. Through access, the American people have begun to expand their expectations of television, because they have begun to understand that more can be expected: more diversity of format, look, and indeed, content.

Interestingly enough, this production quality phenomenon is not, nor ever has been, the case in print media–to return to our earlier comparison. Neither the quality of the paper, nor the type font employed, nor even the printing format has ever had any influence on the print audience's acceptance of the message on the printed page. The print audience was never educated to equate the message with the quality of its medium. Perhaps, then, one of the educational jobs of access television is to educate the audience to separate the messenger from the message.

Another characteristic of the television generation is that they have come to expect the programs they watch to entertain them, so that they watch casually, without paying close attention. These audiences often use television as a background to whatever else they are doing, in the same way that the radio in a car is used while driving.

However, many access programs have quite serious themes; they are not designed to be light fare, but to convey serious concerns. If the viewer finds the program too heavy, he can be expected to look elsewhere on the cable dial or converter touchpad. Our message is that even programs with the heaviest of themes must offer a minimum level of entertainment value or casualness; otherwise, audiences will certainly be lost. The theme can be heavy, but its treatment must be palatable to the casual viewer.

DAY-TO-DAY MANAGEMENT

GENERAL MANAGEMENT THEORY

Since this book is about the management of resources and people, some words about general management theory are appropriate. We would like to begin by describing the two major American management styles, sometimes referred to as the "X" and "Y" styles. We describe them primarily because of their possible impact on the management of volunteers.

In managerial style X, the manager acts as a taskmaster, leaning on or pressing employees to produce more and more, and carefully counting and evaluating every moment of an employee's job time. Using the fear of job loss or demotion for motivation, style X is, sadly, the most prevalent managerial method used in the United States.

This style works well for a limited time and up to a point. When employees working under this managerial style feel personally secure enough in their jobs, they tend to rebel—very quietly and unobtrusively, to be sure—by sabotaging the operation whenever possible and as long as they feel sure that they will not be caught. The classic illustration of this concerns Detroit-made automobiles, and is exemplified by the expression, "Don't buy a car that was made on a Monday or a Friday." Think of what that statement says about the job attitudes of employees just before or just after their days off!

Regardless of the management style of their employer, all employees who have reached the stage where their salaries are adequate for their needs and a few luxuries (although not as much as they perhaps aspire to), begin to work for what is termed "psychic salary." This salary is the "warm and fuzzy feeling," the inner satisfaction that comes from doing a job that is enjoyable. The employees of the taskmaster receive their psychic salary from enjoying the game of trying to outwit their manager, rather than from devoting their energies to their job performance. A job they had applied for and accepted because it seemed to be one that they would enjoy doing becomes a challenge in its undoing.

All other things—such as adequate pay and healthy work environment—being equal, it can be stated that employees come to and stay with a job for the psychic salary that they receive.

In the management style Y, the manager sees his job as that of an expediter or facilitator. He spends his time ensuring that his employees get whatever they need to do their work in the most expeditious way. This manager cuts red tape. He lobbies for more responsibilities and better workplace conditions for his people. He places himself between his employees and upper management, who might wish to sacrifice them in the never-ending search for cost-cutting measures that is one of the hallmarks of our free enterprise system. This manager always has the time to listen when an employee wishes to share a problem about work or about home and family. The manager knows full well that unburdening his problems will help the employee focus all of his attention on job performance.

Under this type of management, employees typically respond by working with great enthusiasm for their manager. The resulting high level of work performance by his people makes the manager less vulnerable to the vagaries and pressures of upper management. The employees under this managerial style get their psychic salary out of the job itself and out of the camaraderie that they share with their manager.

THE MANAGEMENT OF VOLUNTEERS

Managing volunteers is very different from managing employees. As we have mentioned, employees have an economic stake in responding positively to their manager. Volunteers feel no such pressure: they respond as the mood moves them. They show up, or don't, for programs or other duties at the center, directed only by self-motivation.

The access manager will find that many different types of people sign up for access training, and that fully a third of those who have been scheduled do not appear for class. The no-shows will have been diverted by weather conditions, overtime offers at work, illness, chance encounters with members of the opposite sex, and so on. To make matters worse, the no-shows often do not bother to call the center to advise that they will not be there.

Some centers charge a small sum for training, often refundable upon completion. This charge sometimes insures that the prospective trainee will attend regularly and will complete the training. This result benefits both the trainee and the center. However, this type of fee can sometimes discourage participation by the disadvantaged, unless they are permitted to pay for training by volunteering time at the center in lieu of money. Thus, a scholarship with time instead of cash as payment is used as a motivator.

The manager will find that, once trained in the basics of video, most community people will wait for an invitation to take part in access. Few will hit the ground running. Volunteers must be encouraged and invited to become involved as regular program producers. Few will have original program ideas, and some of those who do will be too embarrassed by anticipated rejection to present them.

To start the access program process, the manager should pick a person of substance from among the basic trainees: someone who might be a businessman, a senior government employee, a police officer, or a professional—in short, a motivator. This person should be someone who is accustomed to telling people what to do and offering advice on how to do it without being asked. Offer this person the chance to become a program producer, give him or her a program idea, and provide a list of ten other trainee graduates, complete with their phone numbers. Ask the prospective producer to call these people and request them to be on a production crew. Explain that additional names and phone numbers are available if the people on the first list are entirely or mostly unable to help.

The first response of the prospective producer will either be fearful and negative—which indicates that you have picked the wrong person—or fearful and questioning. "But I don't know anything about being a producer," will be the response. "I will teach you," the manager replies. And, indeed, if the manager has picked the right person, a half-hour one-to-one session will furnish whatever knowledge the producer needs to get started.

Managing volunteers consists of literally pushing them into program production and then relying on the excitement of production to motivate their continuation. One of the biggest problems confronted by the access manager is "transitioning" volunteers from the basic training into being regular access users. Many centers do not effectively address this concern. One may think of the solution as setting a net, or perhaps a catcher's mitt, to catch the graduates of the training and then immediately putting them to work in some capacity in access.

Being a program producer is not the only option open to the access volunteer. Some graduating trainees, who have professional skills in addition to those acquired at the center, can be relatively easily convinced to use their professional skills in access. Graphic designers, carpenters, accountants, and attorneys all have skills that are crucial to access operations at one time or another.

Completed shows must be played back on the channel. Playback skills are easily imparted to volunteers, who immensely enjoy sitting at the equipment and putting the programs on the channel. This system frees staff for other assignments.

In communities with nearby military bases, some volunteers from the military will bring with them program skills and electronic maintenance skills that, under the direction of the center's technician, can be used to keep the access equipment in first-class condition.

On the other side of the coin is the volunteer who comes to the center brimming with program ideas and wanting to discuss them and produce programs. Value that person. You have met someone who will make your job considerably more pleasurable. Shepherd and nurture him or her through the basic training, and begin, in discussions and conferences, to describe the production values that are inherent in good programming. If you give the idea person a crew and guide them through their first production, he or she will go on to provide the primary reason that you find yourself at the access center.

MULTIPLE CENTER MANAGEMENT

In a community large enough to enjoy the benefits of more than one center, there are certain factors at each of the centers which must be addressed in common.

Coordinating Information

There is information that should be commonly held by all of the access center facilities. The easiest and most effective method of addressing information coordination is by using personal computers (PCs). This topic was addressed briefly in Chapter 6 when we spoke about developing record keeping systems. If personal computers are used as coordinating devices, one PC is designated as the main computer and normally located at the primary access facility. This computer—or rather its nonvolitile memory on either hard or floppy disks—with a data base program, becomes the repository of all information about a number of different areas of mutual concern.

PCs at the outlying facilities are then linked to the main computer via telephone modems, so that operators can enter and retrieve information from the central computer. (Modem is the shorthand term for a modulator-demodulator, the device that allows computers to communicate over telephone lines.)

We suggest using the telephone system rather than the cable system to link the computers for two reasons, neither of which have anything to do with loyalty to the cable operator: (1) Modems for communicating by telephone are less expensive and easier to obtain than those for cable systems. (2) Using the cable system limits communications only to computers connected to the cable system; such a linkage prohibits communications with computers on other systems or in other cities.

What kinds of information need to be coordinated among several facilities sharing a data bank?

First are lists of people in the community who are members of the access organization, or who are registered, certified, trained access users. The lists should detail their levels of skill and the training courses that they have taken and satisfactorily completed. It is important to make this information available to all access center facilities because a producer may wish to use one facility that is close to her home and also one that is close to her workplace. Both of these facilities must be able to ascertain what equipment she is certified to use—just the porta-pak, or a studio, or editing facilities.

Each access center needs to be able to enter data to update the information in the central computer file. When a user successfully finishes a training course, that information must be able to be entered into the data bank immediately. All centers will then have available the change in certification level of that user.

The second major need for information coordination between several access centers is so that use of equipment and facilities can be regulated. Suppose, as an example, that a multicenter facility has three access centers. A user arrives at center A with an urgent facilities request. The center's staff recognizes the need, but center A's facilities are checked out or previously reserved for the time the user needs. The access facilitator at center A can make a quick computer check of a centralized equipment data base to determine whether either center B or C has facilities available at the required time. After a quick conference with the user concerning agreed-upon use at another center, center A's facilitators could instantly reserve the facilities for the user. Of course, the same series of events could be accomplished with a couple of phone calls, but without the availability of on-screen information, a zealous facilitator at the other center might inform the caller that all facilities are booked. Thus the other center's facilities are only available to someone who might drop in to that center.

A third reason to coordinate information on computers is to check for access users who have been temporarily barred from using access facilities because they have broken access rules. Clearly, if a user is on probation at center A, he is also on probation at centers B and C. A statement to that effect on the user's central data bank file, giving the nature of the infraction as well as the dates of probation, prevents the devious probationer from merely moving his base of activities to another center after being temporarily barred from the use of one center. Therefore, information on users who are on probation needs to be centralized.

Program scheduling on the channel may be an issue in a multicenter access facility, depending on the scheduling procedures. In some communities all program scheduling is done at a centralized point. At a facility with a computer, the facilitator at the primary access site can immediately enter the date and time of entry into the main program schedule file. This information may become vitally necessary in disputes between producers over whose program was submitted first if identical playback time slots have been requested. Also entered are the name of the completed taped program that has been submitted for cablecast, the length of the program, and the playback date and time requested by the program producer. This computerized system reduces the effort and time needed to shift paperwork between facilities. The Dallas, Texas, access facility found that, without a computer, it was impossible to do adequate and accurate scheduling and shift paperwork among its five centers. Unfortunately, a computer cannot move videotape between centers.

For this extensive use of computers and modems at several sites, we recommend that the access facilities consider more powerful computer equipment than was suggested in Chapter 6; at a minimum, they should go for equipment in the IBM-PC or Apple II class, with hard-disk storage. It is also most helpful if one or more of the access center volunteers is a computer programmer who can write the specific software required for the multicenter access facility. If no programmer is available, then several of the data base computer programs can be adapted to meet the centers' need.

Competition Among Centers

A small amount of competition among centers is healthy, because excellence at one center sets goals for the other centers in a facility to try to exceed. Often, however, competition between access centers may grow so intense that volunteers and staff of one center may develop bad feelings—and worse, show those feelings—toward the users and staff of another center. The access manager, who should keep a finger on the pulses of both the centers, is strongly encouraged to intercede and prevent competition from stretching beyond healthy bounds into fisticuffs.

FIRST AMENDMENT ISSUES AND GATEKEEPER ROLES ____

Because of the Cable Communications Policy Act of 1984 and the raft of court decisions that have followed its passage, and because of other court decisions that are still in progress, access managers live on the edge of an emerging body of communications law. Most of this law derives from the First Amendment to the US Constitution. This amendment, written by Thomas Jefferson and appended to the Constitution in 1791, says in part that "Congress shall make no law . . . abridging the freedom of speech, or of the press; or the right of the people to peaceably assemble." In this modern era the amendment is interpreted to mean that government in general shall uphold the rights of the people to free speech, and, in fact, shall encourage it by its actions.

The access manager must expand her educational horizons and learn a new vocabulary to keep up with the new communications laws. She must learn the meaning of a "First Amendment speaker," the difference between the "content" of free speech and its "conduit," and the definition of a "natural monopoly." These broad, conceptual ideas are normally the preserve of attorneys who specialize in First Amendment or constitutional law. The authors of this text are not attorneys, and therefore the material that follows is clearly opinion. However, our opinions have been shaped by eminent attorneys and by others with vast experience in the field of communications law.

Nevertheless, we face a dilemma in this discussion. We feel that it is important for the manager to be abreast of communications law as it affects cable in general and access in particular. But the reader should be aware that the law is in flux, and so any point that we might emphasize today may very well be obsolete, or its emphasis reversed, tomorrow.

Furthermore, the public policies, based on the First Amendment, that affect access or L/O (local origination) have not had very long to evolve. Certainly print media law has had much more time to develop. Access and L/O are relatively new and the policies that govern them are newer still and are still being evolved. But the manager must live with the law that exists today, and must operate within the evolving framework.

With those caveats and disclaimers, we plunge headlong (and we hope not foolhardily) into a discussion of the ways in which the First Amendment affects access, L/O, and leased access channels.

The Congress of the United States, when it passed the Cable Communications Policy Act of 1984, went on record in support of access on cable television as essential for free expression. A publication of the House of Representatives Committee on Energy and Commerce, *The House Report,* stated in 1984 that access requirements "serve a most significant and compelling governmental interest; promotion of the basic underlying values of the First Amendment."

Thus, by law, access channels can be required of the cable company franchisee by franchising authorities. Services, facilities, and equipment for access can also be required of the franchisee. All funds earmarked for access in franchises in existence as of October 29, 1984, (the date of enactment of the Cable Act) are exempt from the newly fixed franchise fee ceiling of five percent. In new franchises, only funds earmarked for capital costs—that is, costs relating to construction of access facilities and equipment purchase—are exempt from the five percent maximum franchise fee.

The Cable Act significantly altered the previous relationship between the cable operator or the access center and the access program producer. Formerly, the cable company, and often the access center, felt free to act as a gatekeeper, or editor, of material produced for public access channels. But now the cable operator and the center are barred from exercising *any* editorial control—except for obscenity—over access programming.

Responsibility for the editorial content of programs has now shifted directly to the programmers. The act protects the cable company from any liability arising from access programming. Access program producers are now also liable for anything seen or heard on their programs that libels or slanders another person. Thus, the new law puts producers on notice that they are legally responsible for their work. Moreover, they are considered "full First Amendment speakers," a term that refers to a person or entity whose freedom to communicate is unfettered by censorship or by prior restraint.

Cable companys, local governments, or other access management entities that consider, or have considered, censoring the content of access programs will find that the act, by reference to the First Amendment, strictly limits both what may be censored and the procedures that must be followed for censorship. Access has become, in the words of the *House Report*, "the video equivalent of the speaker's soap box or the electronic parallel to the printed leaflet."

The act gives local governments explicit authority to create structural regulations to govern programming on the channels. Structural regulations are those that define methods for accomplishing access, like first-come, first-served procedures or other democratic methods and systems. However, these regulations do not and are not permitted to deal with program content.

It should surprise no one to find that some local governments— and, specifically, members of local governing bodies—and cable companies are threatened by a free flow of information in their communities. These groups are not interested in seeing public access fulfill its intended role under the First Amendment. In these instances, the interests of some segments of the cable industry and those of fearful local governments are the same. Only their reasons are different. The cable industry segment wants control of the access channels for commercial reasons and some local governments fear doing their work in the sunshine of open and free-flowing information.

If these interests combine, the results may be poor or no implementation of access by the cable operator, and little or no enforcement of access provisions by the local government regulatory authorities. The ultimate result is the same: a restriction of the free speech of the citizenry.

Selective enforcement of franchise provisions can further complicate the First Amendment issue. If enforcement is limited only to the provision of educational and municipal access in the franchise agreement, and if public access is minimally enforced or excluded, then the local government can create a situation in which a community's mainstream organizations have local programming capability, but the less powerful and politically unpopular groups do not. These groups must rely on the limited, first-come, first-served resources left over for the access channels. Thus, the appearance of a healthy local programming effort in the community is maintained, while the First Amendment rights of the groups who were promised a voice through access are effectively restricted.

There are other variations on this theme of selective enforcement. Facilities are sometimes made available to one community group, without requiring that group to work with other members of the community. If local government does not intercede by requiring access resources to be made available in a democratic fashion, then the same results can be observed: a restriction of the First Amendment rights of the politically unpopular and powerless.

Often local government is found competing with its citizens for access resources. Government thus vies with the very people and organizations whose First Amendment rights it should be supporting. When government becomes more concerned with doing programming on the system than being regulators of the system, it comes into direct conflict with its responsibility to protect the First Amendment rights of its citizens. This attitude of local government always results in selective enforcement of the access franchise provisions.

Local government is also involved in access through its role in starting and incorporating nonprofit organizations to manage access resources on behalf of the public. This involvement may extend to monitoring the budget of the access corporation, appointing access board members, or allocating part of the government's franchise fee to support public access.

Such deep involvement by local government implies that some possible First Amendment considerations should be examined. As we have said, the First Amendment prohibits government from inhibiting an individual's free speech, and it also gives the government the responsibility of upholding the rights of individuals to speak in a public forum. In the access context, a cable system is a public communications forum, and public access channels are required in their franchises to protect people's First Amendment rights to speak in that forum.

Another potential conflict of interest might arise at the government level: while protecting the public access rights of one speaker or a group of speakers, government might restrict the rights of others by exclusion. The recognition of this possible conflict has been a major impetus in the creation of nonprofit access corporations that take the responsibility for the role of access outside of the jurisdiction of local government and the cable company. This relationship is properly described as an arms-length relationship: not adversarial, but business-like and supportive.

Even with government influence supposedly limited by law to the structural regulations mentioned earlier, local government's recent role in access development has been so extensive that real potential for abuse of the First Amendment exists—and not just theoretically. In any gathering of access people, you will shortly hear discussions of instances in which government has restricted the content or diversity of access programming.

The first problem area is the censorship, and the second is the failure of government to evenly enforce access provisions. Both of these issues have just been detailed. The third, and perhaps the most pervasive issue, is the government influence of access organizations by political and budgetary pressure.

Where the local government appoints the access board, oversees the budget, allocates a portion of the franchise fee, and sometimes staffs and administers access, very little imagination is needed to see the possibility of inappropriate government control. In such instances, there is a danger that through partisan politics, oversensitivity to criticism, or arbitrary perceptions of tasteful and appropriate programming, local governments will discriminate when allocating access resources.

Political pressure may even include attempts at outright censorship of a particular program if a local bureaucrat thinks that he can get away with such an action. However, the issue will rarely be that well defined and clear cut. Nor will it be that easy to bring the offense to public notice and censure the official.

For example, in the community where one of the authors of this book is the executive director of the access corporation, the municipal government created a nonprofit access organization. The corporation was free to manage access until it was about a year-and-a-half old, when nearly fifty solid programming efforts—series programs and one-shots from varied segments of the community—had been cablecast. Suddenly the access corporation came under fire from one or two council members and the local paper on the vague charge of being "unresponsive" to the community. Quickly there was a call to revoke the proclamation that provided forty percent of the five percent fee for access funding. After a nonvoting work meeting, where the concerns of council members were aired and defended, the situation was referred to the city manager and city staff for a so-called study.

In another example in what we will call city X, local government officials took umbrage at a series of programs produced by a local gay rights group. The first inquiry focused on why such a program was permitted on the access channel in the first place. The response from city X's access management organization described the democratic, first-come, first-served principles of the access system. Next, the city administration officials wondered aloud whether the nonprofit organization should change its nondiscriminatory policy so that "unsavory" groups could be excluded. The access management of course explained the First Amendment implications of that type of censorship. They added that such a rule change could generate a legal action for restricting free speech. Not long after, the government officials of city X began to wonder publicly whether city X's access management organization should continue to receive its present level of funding from the city from the cable franchise fees. Finally, the city X government called for an evaluation of the access organization that was directly related to its funding.

Budgetary pressure is difficult to resist; it is pervasive and unending. And it is totally unrealistic to expect access managers to resist this type of government pressure for very long. In the long run, such pressure will create access managers who have learned to give the government what it wants.

In a third example, in city Y, the cable system was owned by the city and operated by a city utility commission. After a series of programs in obvious poor taste appeared on both the access and L/O channels, city officials and some members of the community were angered. Shortly after, the program coordinator was laid off and the access center was closed. The reason given for closing the center was security: a video monitor had apparently been stolen from the center (which is rather like closing a hospital because a bedpan was stolen).

Following these events, a grassroots movement founded a community access committee, which had both program advisory and administrative functions given to it by the city. The access center studio was reopened and the coordinator reinstated. Although it is encouraging that grassroots efforts can have a positive impact on access, one wonders what the effects of the government's action described will be on the center and its coordinator. The center may continue to adhere to the First Amendment goals of access, but once burned. . .

What the community needs from responsive, responsible local government, is help in setting up the nonprofit organization to manage access. Then government needs to imbue that organization with as much autonomy as possible. And then it needs to back off.

Access budgets, as a public trust, should certainly be held to actuarial accountability. But beyond that, government administration of access programming places government and the free flow of ideas too close together and should be scrupulously avoided. In close proximity, the two are not consistent with First Amendment principles. Even when the most enlightened and progressive officials are in power, the potential for abuse is too great. Furthermore, when the organizational structure of access is planned, it should be designed to endure different political climates. Planning based on personalities is always unwise.

Restrictions on government's involvement in media are a part of this country's political tradition. Those restrictions were put in place as the nation was formed and they remain sturdy. Public access is an expression of that tradition, and it requires careful consideration if access is to serve its purpose of ensuring a more democratic flow of information within the community.

City governments are not the only entities that have unwisely abused their gatekeeping roles. For example, nonprofit access management corporation A was a new access corporation that had been in business since about 1982. This corporation decided that they were not going to permit any religious group, or any group that happened to meet in religious buildings, to use the access channel. They took this decision because they were seriously concerned about the separation of church and state. This access organization received funding from a franchise granted by a unit of local government, and they therefore felt that it was inappropriate to permit church groups to have access to the channel. This is, of course, a clear abuse of the gatekeeping role. Access channels are to be operated as public forums. In a public forum, the gatekeeper can only restrict time, place, and manner; the gatekeeper cannot restrict speech or content.

In another example, cable company Z had developed a format for some of its more creative access producers who wanted to do comedy and political satire. The format was not unlike "Saturday Night Live," and it used the skit style of presentation. A community access producer devised a seven-minute piece that satirized sex on television. The piece was submitted to the cable company program director for inclusion in the program.

The seven-minute satire did not include any four-letter words, nor did it include nudity. Although it did use some double entendres, the material was quite mild compared to some of the skits on "Saturday Night Live." However, the cable company refused to permit the program segment to be cablecast. The cable company was clearly censoring a program, and thus abusing its role as gatekeeper.

AUDIENCE MEASUREMENT IN A NARROWCASTING ENVIRONMENT

AUDIENCE MEASUREMENT IN BROADCASTING

In television broadcasting, audience measurement is performed to determine the maximum price that can be charged to a commercial sponsor for a spot announcement, for commercial sponsorship of a program segment, or for an entire program. These prices form the basis of a station's or a network's "rate card," which lists the prices charged for spots, segments, and programs at various times during the program day. The card also outlines the discounts that an advertiser receives for multiple dated advertising, that is, the same ad at different times of day, or different days. The rate card is updated periodically, following "sweep weeks," to reflect increases or decreases in the size of the viewing audience.

The measurement entails assessing two audience parameters, known as "share" and "rating." A number is assigned to each by the rating service. A rating number is a percentage of the total number of television households in the United States viewing a program. A share number is the percentage figure that the audience rating had in relation to all of the other television programs on the air during the same time period.

For example, to determine a rating for a TV program, one must start with the total number of TV homes in the United States. This was 73 million in 1981. If a program earns a "20" rating, that means that 14.6 million homes, or 20 percent of the total, saw the program, which in turn translates to an audience of 33.5 million people. The audience figure is arrived at by multiplying the estimated number of the homes with television by a standard conversion factor of 2.3 people per home. The rating figure counts the total audience that remains tuned to the program for more than five minutes.

A share is the percentage figure ("share of audience") that the rating had relative to viewers of all of the other programs being broadcast at that same time period. For example, if the estimated 33.5 million viewers who watched the program above represented 30% of the total audience watching television during that specific time period, then the program would have a 30 share, or 30% share of audience.

These measurements are mainly gathered by two nationwide collection services, Arbitron and A.C. Nielsen Company, and are then used in a formula that includes population demographics to determine rates for specific stations.

AUDIENCE MEASUREMENT IN ACCESS

Measuring public access audiences—or, indeed the habits of cable television viewers—is more accurately termed audience ascertainment. It is performed to determine the degree of program acceptance, the demographics in terms of who is watching, and the community's needs and desires for programming.

Measuring cable audiences is done through random sample mail surveys or through telephone surveys in communities that have cable television. The surveys may be done by the access corporation, the cable company, or cooperatively by both, since audience sampling is expensive. Surveys are very labor-intensive, and it would be far too expensive for access to use the collection services that are employed by broadcasters.

Often the cablecasters will enlist the aid of the marketing department at the local college or university. This cooperation provides an unbiased sampling organization for the access corporation or cable company, and gives valuable job experience in sampling techniques to the college students taking the marketing course.

Audience Surveys

When conducting surveys, it is important to use terminology that is familiar to the viewers, rather than the terminology that grows up in an access center or a cable company. Consistency in surveying is also very important. If the same questions are not asked of everyone surveyed, then the survey has little value.

For the broadest possible picture, it is important to survey many constituencies. Among those chosen might be community groups, users of other access centers, community leaders, subscribers to the system—or, separately, nonsubscribers—access producers, access board members, and many others.

Survey respondents might be asked to include a description of their occupation with their replies. If the respondent indicates a skill that would be valuable to the access center, then it would be valid to ask the respondent if he or she would consider volunteering at the access center to help broaden its skills base.

If it is feasible and if the cable company permits it, consider mailing the survey with a cable service billing or with the monthly program guide. This approach saves postal fees and typically yields a forty to sixty percent return rate on the questionnaires.

It is often valuable for the access center to provide an incentive to people who return a completed mail survey. The incentive might be a gift supplied by a local merchant and sent to a randomly picked selection of those who completed the survey; often the gift is a bumper sticker, poster, or T-shirt that prominently displays the access channel logo.

If the cable system has a large enough interactive component, that is, a two-way cable system where the viewer may interact, consider using it for surveys. Conducting surveys at local shopping malls during the busiest shopping times is also valuable.

A SURVEY OF COMMUNITY ACCESS VIEWERS

As an example, we now present the essence of a survey report submitted at the conclusion of a survey performed between March 13 and 20, 1985. It was prepared by a team of university researchers for Professor Frank R. Jamison, the head of Media Services, Western Michigan University, Kalamazoo. Professor Jamison is presently a member of the National and State Board of Directors of the National Federation of Local Cable Programmers (NFLCP), and the findings of this viewership survey were presented in a major paper at the NFLCP annual convention at Boston, Massachusetts, in July 1985.

The Survey Report Contents

The survey was designed to gather information on the viewership of community access programming; specifically, it was intended to study whether access has any impact on cable television viewing. That is to say, if community access were discontinued, what effect—if any—would it have on cable subscriptions?

The project was initially focused on Kalamazoo, Michigan, where the dominant cable operator is Fetzer Cable Vision. The survey was then compared to similar national surveys. The inquiry was undertaken by a telephone survey of cable subscribers in the Kalamazoo metropolitan area. In addition, certain survey questions were evaluated to provide a better understanding of the characteristics of community access television viewers. The results of this research were then compared to other national surveys to attempt to define a general trend.

The survey gives the percentages of responses to each survey question for both users and nonusers of community access television. When the responses to the survey were tallied and correlated, the results showed that community access does play a small but significant role in the decisions of the residents of Kalamazoo to subscribe to cable television.

Methodology of the Survey

The survey was designed to be used as a technique for predicting the market behavior of cable subscribers. Within the parameters of the project, the survey had to be technically accurate in every sense so that it might be adapted for use at the national level. To aid in the survey design, the research team interviewed experts in both the cable television and market research fields, who gave advice on survey instrument design and selection of sample size. The experts chosen were from the marketing research faculty at Western Michigan University, the communications faculty at Michigan State University, and professionals at the Kalamazoo Community Information System for Human Services; the Kalamazoo City Cable Administrator and Community Access Center Director also gave valuable advice.

The telephone survey method was chosen for its efficiency and cost advantages. Telephoning was determined to be efficient because it can be conducted more quickly than face-to-face interviews or mail surveys; it also reduces the potential for bias because there is less social interaction between the interviewer and the respondent. The information elicited by the survey focused on the respondent's behaviors, attitudes, and characteristics.

There are inherent limitations in a telephone survey. Some respondents are unable to answer all the questions because they can't remember. Some bias may result from residents with unlisted telephone numbers or no telephones. Although these factors can limit the validity of the survey, they can be controlled by proper survey design.

Question–Content Decisions

The first section of the survey questionnaire contained respondent identification data: name, phone number, date and time of the call. Interviewers were thus able to tally all the calls made and avoid repeating calls to those already surveyed.

The survey's introduction was designed to enlist the respondent's help, gain cooperation, and establish rapport. It introduced the interviewer as a student, which tended to promote a higher response rate. When introducing themselves, the interviewers requested a male respondent, because previous studies have shown that this necessary category of respondent is more difficult to fill.

The body of the survey flowed from more general questions to specific queries about cable and community access programming. Most of the questions were multiple choice, since this type of question tends to reduce interviewer bias and is easier and faster to administer. The multiple choice technique also reduces the cost and time involved in data processing.

Open-ended questions were used only when necessary. Question 5 (see Figure 12.1) was left open to see if the respondent could name the access channel numbers and if they could distinguish between community access and PBS programming. Question 9 was open-ended because it was important to know specifically what programs were watched. Question 11, which was very important to the study, had to be left open to determine how much less, if any, the respondent would pay if community access were not available. Question 12 was left open because pretesting indicated that a respondent is more likely to provide his or her date of birth than numerical age.

The survey questions are relatively basic requests for cable television and demographic information. General research quidelines for data collection were followed when the questions were worded. The guidelines included the use of simple, clear words and the avoidance of leading questions. To help insure that the questions were asked in a logical sequence and to help visualize the structure of the questionnaire, a flow chart plan was developed, as shown in Figure 12.2.

The questionnaire was pretested using a convenience sampling procedure—that is, calling thirty respondents randomly selected from the Kalamazoo telephone directory. This was done to ensure that the questions flowed properly and were understandable. After the pretest was concluded, areas of confusion were revised and the final draft was ready to be provided to the interviewers.

Planning the Interviews

When the questionnaire was completed, the researchers developed a timetable which is given in Figure 12.3. It was determined that approximately 384 completed surveys were needed for a valid sampling. Volunteers from the university's honorary radio/TV fraternity and from the community access center were asked to be interviewers. In addition, an

Phone # _____   Respondent's Surname: _____
                                                              (from phone book - don't ask)
    Callbacks

            1   2   3   4   5   6

Date of connected call: _____ March 85

                 Time: _____ a.m.   p.m.

Status: Completed _____   Verified _____
         Disconnected _____
         Refused _____

Hello, my name is _____. I'm a student calling from
the Media Services Department at Western Michigan University. We are conducting
a survey on television viewing in the Kalamazoo area. (If female answers)
For statistical purposes we would prefer a male respondent. Is the man of the
house available? (If yes, repeat opening for man)
(If no, say "Your opinions are also important to us....")

        Would you take a few minutes to answer some questions?

            1. Yes (Continue)
            2. No (Thank them and say good-bye)

    Do you subscribe to Fetzer CableVision Services?

            1. Yes, I have cable (Continue)
            2. No. Why not? (Do not read list)
                1. Cost
                2. Not interested in cable programming
                3. Not available
                4. Other _____
                (Skip to sentence above question #12 and continue)

    1. Why do you subscribe to cable television? (Do not read list)

            1. Better reception
            2. More channels
            3. Premium channels
            4. Community access channels
            5. Other _____

    2. What premium cable services, for example, HBO, Movie channel, Disney
        do you currently subscribe to? (Do not read list)

            1. HBO (Home Box Office)
            2. Showtime
            3. Movie Channel
            4. Cinimax
            5. Pass (Pro Am Sports System)
            6. None
            7. Other _____

FIGURE 12.1 Kalamazoo survey instrument. Courtesy of the NFLCP files.

-2-

3. How many times do you select "pay per view" services, called Premier
    27 here in Kalamazoo? (Do not read list)

    1. Never
    2. Once or twice a month
    3. Three or more times a month
    4. Don't know

4. Approximately how much money do you spend monthly for cable television
    services in your home? _____ (Do not read list)

    1. Less than $15.00
    2. $15.00 to $30.00
    3. $30.00 to $45.00
    4. More than $45.00
    5. Don't know

5. Are you aware that your cable system has community access programming?

    1. Yes Can you name the channel numbers? _____
    2. No

6. Channels 5, 7, and 9 are community access channels. Have you ever
    watched these channels?

    1. Yes (Continue)
    2. No (Skip to sentence above question #12 and continue)
    3. Don't know (Skip to sentence above question #12 and continue)

7. How often have you watched these community access cable channels in
    the last two weeks?

    1. Not at all
    2. 1 to 2 times
    3. 3 to 4 times
    4. 5 or more times

8. We are interested in the programs you have watched on the community
    access channels. Have you watched..... (Read list)

Yes	No	Don't know	
1	2	3	Government programs
1	2	3	Local Sports Events
1	2	3	Arts and Entertainment
1	2	3	Health and Wellness
1	2	3	Religious
1	2	3	Political/Public Affairs
1	2	3	Children's programming

9. Can you think of any particular access programs that you have seen
    in the past two weeks? (Record all responses)

    _____

    _____

-3-

10. How important was the presence of community access programming in your decision to subscribe to cable? (Read list)

    1. Not at all
    2. Somewhat
    3. Important
    4. Very important

11. Would you be willing to pay the same for basic cable service if community access channels were not available to you?

    1. Yes
    2. No - How much less $_____
    3. Don't know

The following questions are for classification purposes only and will be kept strictly confidential.

12. What was the year of your birth? _____

13. What is your marital status?

    1. Single
    2. Married

14. What education level have you completed? (Do not read list)

    1. High school graduate or less
    2. Some college
    3. College graduate
    4. Refused

15. Are you currently employed?

    1. Yes (Continue)
    2. No (Go to question #17)
    3. Retired (Go to question #17)
    4. Student (Continue)

16. Would that be....

    1. Full time
    2. Part time

17. Approximately what is your estimated total household income for this year? (Read list)

    1. $15,000 or below
    2. $15-30,000
    3. $30-45,000
    4. Over $45,000
    5. Refused

Those are all of the questions I have. Thank you very much for your time.

Note sex of the respondent:

    1. Female
    2. Male

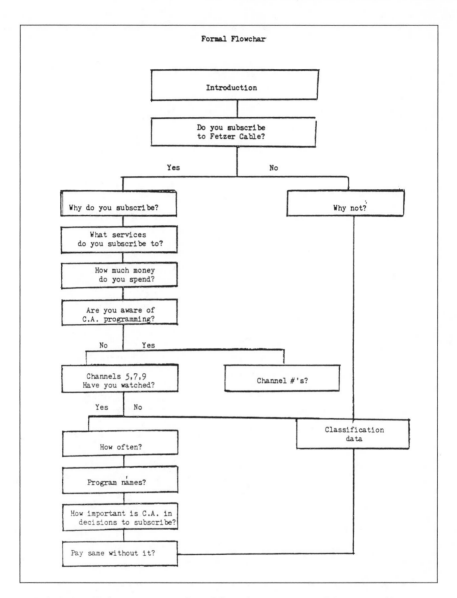

FIGURE 12.2 Kalamazoo survey formal flow chart. Courtesy of the NFLCP files.

```
                                   Time Schedule

        Phase of Study                  Time Involved

        Draft pretest                    6 weeks

        Select pretest sample            1 day

        Select pretest interviewers      2 weeks

        Training material                2 hours

        Pretest interviewing             3 hours

        Evaluate Pretest                 3 days

        Finalize forms                   1 week

        Select interviewers              2 weeks

        Train interviewers               2 hours

        Interviewing                    18 hours

        Analyze results                  2 hours

        Report                           2 weeks
```

FIGURE 12.3   Kalamazoo survey timetable. Courtesy of the NFLCP files.

alphanumeric request was placed on the community access channels' bulletin boards. Interviewers were given instructions for recording responses, and these are listed in Figure 12.4.

The telephoning was performed at a central location so that the researchers could have close control over the survey. Out of a total of 1,326 telephone calls, 415 surveys were completed.

The project was implemented on a limited budget. Telephone connection costs, printing costs, and the data processing fee were all funded by the project's client, Professor Jamison.

Sample Design

The sample size was determined using the statistical formula shown in Figure 12.5. The size of the Kalamazoo population is not mentioned in the calculations for two reasons: First, the related population from which the sample was drawn is large enough that it becomes statistically unimportant. Second, the variance of the population, not its size, determines the sample size. Variance means the degree of dispersion from the average (mean) of the population.

Convenience sampling procedures were selected because of budget constraints. Respondents were selected from a current telephone book by dividing the book into quarter sections; selecting a name and phone number from the first quarter section and then skipping to the next column and selecting the name that falls in the next quarter of the page, and so on. This sampling procedure was recommended because it is easy to administer and inexpensive.

Data Processing

The researchers were taught how to convert raw data from the survey into a form that could be manipulated by a computer. Because most of the questions were closed-ended—that is, multiple choice—Scan-Tron recording sheets were used to transfer survey answers to a computer program developed for the survey by a graduate assistant in the marketing department at Western Michigan University. The program allowed for cross-tabulation of survey questions, a computer print-out of information, and a data file.

SURVEY FINDINGS

The results of the survey were given in percentages based on the total number of people who answered each question. The results of the 415 com-

General Interviewing Instructions

Brief explanation of market research and purpose of study.

Rule 1

Be pleasant and courteous.

Rule 2

Always ask each question exactly as worded.

Rule 3

Ask questions in the order they appear on the questionnaire.

Rule 4

Do not read the listed answer choices for a question, unless the answers are included in the question or the instruction "Read List" appears on the questionnaire. Never read "Don't Know" or "refused."

Rule 5

On open-end questions, always record the respondent's answer verbatim.

Rule 6

Get an answer to every question that should be asked, or be sure to mark "Don't Know" if respondent cannot answer.

Rule 7

As soon as you complete interview, check the questionnaire carefully for omissions.

Rule 8

Never suggest anything to a respondent or lead them into a answer.

FIGURE 12.4 Kalamazoo survey interviewing instructions. Courtesy of the NFLCP files.

Appendix E

Sample Design

I.   Population:

    A.   Elements - Fetzer Cable subscribers.

    B.   Sampling Units - Fetzer Cable subscribers in Kalamazoo.

    C.   Extent - Kalamazoo, Michigan, and surrounding area.

    D.   Time - March 13-20, 1985.

II. Sampling Frame - List of Kalamazoo residents from phone book.

III. Sample Size

    $p = .04$ (success)

    $q = .96$ (failure)

    $z = 2$ for 95% confidence interval

    n = sample size

$$n = \frac{z^2 pq}{E^2} \qquad\qquad n = \frac{2^2(.04)(.96)}{(.02)^2} = \frac{.1536}{.0004} = 384$$

    Where: p equals the expected percent of individuals who would be willing to pay less for basic cable service without community access programming.

              q equals 1 - p

              z is the confidence interval, which is set at 95%. This means under repeated sampling, we would expect the true population mean to fall within such intervals 95 samples out of 100.

IV. Sampling Procedure:

        Convenience

V.   Method for determining accuracy:

    Standard Error $Sp = \sqrt{\frac{pq}{n}}$

$$Sp = \sqrt{\frac{(.04)(.96)}{384}} = \sqrt{\frac{.0384}{384}} = .0001$$

FIGURE 12.5   Kalamazoo survey sample design. Courtesy of the NFLCP files.

pleted surveys were analyzed and separated into three categories: general cable data, community access data, and demographics.

General Cable Data

The results of the general questions broke down as follows:

Do you subscribe to Fetzer Cable Vision services?
1. Yes   61%
2. No   39%

Why not?
1. Cost   13%
2. Not interested in cable   30%
3. Not available   31%
4. Other   26%

1. Why do you subscribe to cable television?
   1. Better reception   16%
   2. More channels   67%
   3. Premium channels   5%
   4. Community access   .2%
   5. Other   23%

2. What premium cable services do you currently subscribe to?
   1. HBO (Home Box Office)   35%
   2. Showtime   2%
   3. Movie Channel   0%
   4. Cinemax   10%
   5. Pass   4%
   6. Disney   4%
   7. None   54%

3. How many times do you select "pay per view" services, called Premier 27, here in Kalamazoo?
   1. Never   90%
   2. Once or twice a month   8%
   3. Three or more times a month   1.5%
   4. Don't know   18%

4. Approximately how much do you spend monthly for cable television services in your home?
   1. Less than $15.00   44%
   2. $15.01 to $30.00   35%
   3. $30.01 to $45.00   2%
   4. More than $45.00   1%
   5. Don't know   18%

Fetzer Cable Vision is the dominant cable service in the Kalamazoo area; 61% of those surveyed subscribed to its service. The predominant reason for subscribing was the added channels that cable offers the viewer. Among subscribers, the majority (54%) did not subscribe to any of the premium services, preferring only the basic service. Among the respondents who did not subscribe to Fetzer Cable Vision, the primary reason was that the service was not yet available to them. Others stated that they were not interested in cable programming.

Premium service subscriptions are reflected in question 4, which asked the respondent's monthly payment for cable. The answer most frequently given was $15 or less. In question 3, the pay-per-view service captured a 10% viewership; half of those respondents reporting viewing the service only once or twice a month. HBO was the most popular premium service with 35%.

Community Access Data

5. Are you aware that your cable system has community access programming?
   1. Yes   86%
   2. No   14%
   Can you name the channel numbers?
   1. Channel 5   17%
   2. Channel 7   51%
   3. Channel 9   41%
   4. Other channel named   7%

6. Channels 5, 7, 9 are community access channels. Have you ever watched these channels?
   1. Yes   76%
   2. No   23%
   3. Don't know   1%

Questions 7 through 11 pertain only to people who have watched access programs in the past.

7. How often have you watched these channels in the last two weeks?
   1. Not at all   38%
   2. 1 to 2 times   32%
   3. 3 to 4 times   17%
   4. 5 or more times   13%

8. We are interested in the programs you have watched on the community access channels. Have you watched:
   1. Government programs   44%
   2. Local sports events   52%

3. Arts and entertainment   42%
4. Health and wellness   39%
5. Religious   28%
6. Political/public affairs   46%
7. Children's programming   22%

9. Can you think of any particular access programs that you have seen in the past two weeks?
   1. Kalamazoo City Council   24%
   2. Portage City Council   4%
   3. Basketball   9%
   4. Hockey   2%
   5. Aerobics   6%
   6. Under the Rainbow   2%
   7. Belly Dancing   1%
   8. Others   32%

10. How important was the presence of community access programming in your decision to subscribe to cable?
    1. Not at all   74%
    2. Somewhat   19%
    3. Important   5%
    4. Very important   2%

11. Would you be willing to pay the same for basic cable service if community access were not available to you?
    1. Yes   81%
    2. No   13%
    3. Don't know   6%
    If no, how much less?
    1. $0–2.00   27%
    2. $2.01–$4.00   15%
    3. $4.01–$6.00   19%
    4. $6.01–$8.00   2%
    5. $8.01–$10.00   11%
    6. Would drop cable 5%
    7. Unsure   21%

Because most viewers (67%) decided to subscribe only to the basic service, it is considered possible that community access channels helped to make the basic cable package more attractive to viewers.

Kalamazoo seems to have done a good job of making its citizens aware of community access programming, because the vast majority of those surveyed (86%) were aware that their cable service included the access channels.

Of all of the Fetzer Cable Vision subscribers, 76% had watched the access channels at one time or another. To verify that the respondent was an access watcher, and not confusing public access programs with those on PBS, the survey asked if the viewer could give numbers of the channels that had community access programming. Over half (51%) could identify channel 7, and 41% named channel 9. Channel 5 had only recently begun access programming, which may explain why only 17% recognized it.

To reconfirm that viewers were aware of access services, they were asked later in the survey if they could name any specific access programs that they had viewed. Six access programs were listed five or more times: the Kalamazoo City Commission meetings were named most frequently. Other access programs that were frequently listed were high school and college basketball and aerobics.

To consider the respondent a verified access viewer, he or she must have stated that access programming was viewed, identified one or more access channels, and/or named a specific access program. Based on these criteria, 61% of all Fetzer Cable Vision subscribers were verified access viewers. This indicates that access is effective in communicating with its community.

Question 7 dealt with whether the verified viewers were regular access watchers or tuned-in occasionally for special programs. This was resolved by inquiring how often the viewer had watched access programming during the two weeks preceding the survey. The question found that 62% had viewed access one to two times. This figure is almost identical to the percentage of verified viewers, and supports the premise that verified viewers watch on a semiregular basis.

Question 8 reflects the types of programs that verified viewers watched most frequently. The most popular programs were local sports events, which 52% of the verified viewers watched on a semiregular basis. Following in descending order, were political/public affairs programs (46%), government programs (44%), and health and wellness programs (39%).

The responses to question 10 related directly to the main purpose of the survey. It asked respondents how important the presence of community access programming was in their decision to subscribe to cable. Twenty-six percent of the respondents described community access programming as somewhat important, important, or very important in making this decision. This finding is especially useful for helping cable companies understand that community access programming plays a role in the marketing of cable subscriptions.

To explore this aspect even further, question 11 asks respondents whether they would pay the same for cable if community access were not available. Thirteen percent responded that they would want to pay less for basic cable service if community access were discontinued. This figure is much higher than presurvey predictions and indicates that community access has a monetary value to Fetzer Cable Vision, as well as to viewers.

Demographics

The following questions contain demographics of all persons who participated in the survey.

12. What was the year of your birth?
    1. 1967–1985   16%
    2. 1955–1966   6%
    3. 1943–1954   23%
    4. 1931–1942   28%
    5. 1919–1930   14%
    6. 1919–below   13%

13. What is your marital status?
    1. Single   41%
    2. Married   59%

14. What education level have you completed?
    1. High school graduate or less   39%
    2. Some college   24%
    3. College graduate   36%
    4. Refused   1%

15. Are you currently employed?
    1. Yes   61%
    2. No   28%
    3. Retired   11%

16. Would that be:
    1. Full time   82%
    2. Part time   18%

17. Are you currently a student?
    1. Yes   9%
    2. No   91%

18. Approximately what is your estimated total household income for this year?
    1. $15,000 or below   19%
    2. $15,000–$30,000   28%
    3. $30,000–$45,000   20%
    4. Over $45,000   15%
    5. Refused   18%

Sex of respondents
    1. Female   53%
    2. Male   47%

To be fully meaningful, the survey had to determine the type of people who had participated. To do so, questions were asked about the respondent's age, marital status, education, employment, income, and sex.

Based on the leading categories from each set of questions, the survey found that the composite viewer is between the ages of thirty-one and forty-five, married, a high school graduate or less, employed full time, with an income between $15,000 and $30,000. The survey found that 53% were female and 47% were male.

Cross-Referencing

Cross-referencing of responses was done to derive more information about the average access viewer. Statistics on access viewers were cross-referenced with the most important cable-related responses and with the personal characteristics of access viewers. The results indicate several differences in access viewers in terms of the type of program viewed and sex, income, and education level.

The first cross-reference analyzed whether family income was related to the amount spent for cable or community access services. Table 12.1 indicates a definite correlation between the responses to questions 4 and 18. Taken together, the findings indicate that as family income increases, so does the amount spent monthly for cable services. Cable viewers who had a household income of over $45,000 made up the largest percentage of those who spent more than $30 per month for cable services. Income, as might be predicted, played a significant role in subscribers' choice of more expensive cable packages.

The next cross-reference compared verified community access viewers and their stated personal characteristics; this was studied to determine if access viewers had different characteristics from nonwatchers. Table 12.2 documents the comparisons. The largest category of verified access viewers

TABLE 12.1 • CROSS-REFERENCING OF SURVEY QUESTION 4 (MONEY SPENT ON CABLE) WITH SURVEY QUESTION 18 (FAMILY INCOME)

Income	Money Spent Monthly on Cable (Percent)				
	0–$15	*$15–$30*	*$30–$45*	*Over $45*	*Don't know*
$15,000 or below	8	5	0	0	2
$15,001–$30,000	13	9	0.5	0	4
$30,001–$45,000	7	11	0.7	0.2	4
Over $45,000	7	5	1.2	0.2	4
Refused	9	4	0	0.2	5

Source: Courtesy of NFLCP files.

TABLE 12.2 • CROSS-REFERENCING OF SURVEY QUESTION 6 (DO THEY WATCH ACCESS) WITH SURVEY QUESTIONS 12–17 (DEMOGRAPHICS)

	Do They Watch Access (Percent)		
	Yes	*No*	*Don't know*
Employed			
Yes	49	14	1
No	20	6	2
Retired	6	4	1
Education			
High school or less	24	10	5
Some college	20	5	0.2
College graduate	31	8	0.2
Refused	0.2	0.2	0
Birth			
1967–1985	10	6	0.2
1955–1966	4	2	0
1943–1954	13	5	0.2
1931–1942	24	4	0.2
1919–1930	13	2	0.2
1918–below	11	4	0
Income			
$15,000 or below	10	6	0
$15,000–$30,000	21	6	0
$30,000–$45,000	19	5	0.3
Over $45,000	15	2	0.5
Refused	13	5	0.3
Work full or part time			
Full time	65	17	0.4
Part time	10	5	0
Marital status			
Single	28	11	0
Married	47	12	1

Source: Courtesy of the NFLCP files.

had a college education (31%), and the biggest percentage of nonwatchers (10%) were high school graduates or below. The largest group of access viewers in the survey was born between 1931 and 1942, as opposed to nonaccess viewers, who were born either between 1943 and 1954 or between 1967 and 1985. Thus the composite access viewer is a more mature and educated individual. The typical access viewer is also employed full time, married, and has an income of between $15,000 and $30,000.

The next cross-reference compared viewer demographics with the importance they gave to access in their decisions to subscribe to cable. This correlation reveals differences among viewers who ascribed "some" importance to access and those who thought it was "not at all important" in their decision to subscribe to cable. Tables 12.3 and 12.4 present the results. Those who found access unimportant in their subscription decision were mainly between the ages of forty-three and fifty-four, married, college graduates, female, employed full time, and with an income of $15,000 to $30,000. The dominant characteristics of those who found access to be of some importance in their decision to subscribe to cable varies significantly. They were between the ages of forty-three and fifty-four, single, high school or less education, male, employed full time, and with an income of $30,000 to $45,000.

The next cross-referenced responses studied the personal demographics of the viewer who would be willing to pay the same for basic cable without access, as indicated in Table 12.5. The survey found that viewers who would pay the same for cable service without access tended to have the same characteristics as those who considered access programming to have a monetary value. The common characteristics were married, college graduate, employed full time, and born between 1931 and 1943. One variation was family income: those who would pay the same had a larger income—$30,000 to $45,000—than those who placed a monetary value on community access channels, who earned $15,000 to $30,000.

The final cross-reference compared the importance that access played in the decision to subscribe with the willingness to pay the same for basic service without access. The chart in Table 12.6 shows that 13% of the viewers were not willing to pay the same without access; half of those 13% also said that community access channels played a role in their decision to subscribe to cable television. The viewers in this category are the ones believed to place the most value on community access programming.

NATIONAL SURVEY COMPARISONS

Surveys similar to Kalamazoo's were sought and collected from around the nation. An effort was made to collect all the existing data of relevance to community access programming. Cable surveys from the following cities were used in the national correlation: Fayetteville, Arkansas; Tucson, Arizona; Inglewood, California; Longwood, Colorado; Jacksonville, Florida; Muscatine, Iowa; Evanston, Illinois; Bloomington and Indianapolis, Indiana; Ann Arbor, E. Lansing, Grand Rapids, Kalamazoo and Portland, Michigan; Gorham and Mankato, Minnesota; Tulsa, Oklahoma; Portland, Oregon; Berks County and Erie, Pennsylvania; North Texas, Texas; Stoughton and Whitehall, Wisconsin.

TABLE 12.3 ● **CROSS-REFERENCING OF QUESTION 10 (IMPORTANCE OF ACCESS) WITH QUESTIONS 12–17 (DEMOGRAPHICS)**

	Importance of Access (Percent in Parentheses)			
	Not at all	*Somewhat*	*Important*	*Very imporant*
Year of birth				
1967–1985	29 (9.4)	5 (1.6)	6 (1.9)	1 (0.3)
1955–1966	8 (2.6)	10 (3.2)	0	0
1943–1954	40 (12.9)	10 (3.2)	2 (0.6)	1 (0.3)
1931–1942	74 (23.9)	19 (6.1)	3 (1)	2 (0.6)
1919–1930	44 (14.2)	8 (2.6)	2 (0.6)	0
1918–below	33 (10.7)	7 (2.3)	4 (1.3)	1 (0.3)
Marital status				
Single	72 (23.7)	28 (9.2)	10 (3.3)	3 (1)
Married	152 (50)	30 (9.9)	6 (2)	2 (0.7)
Education level				
High school, less	72 (23.7)	22 (7.2)	5 (1.6)	1 (0.3)
Some college	59 (19.4)	15 (4.9)	5 (1.6)	2 (0.7)
College graduate	93 (30.6)	21 (6.9)	6 (2)	2 (0.7)
Refused	1 (0.3)	0	0	0
Currently employed				
Yes	147 (48.5)	37 (12.2)	9 (3)	4 (1.3)
No	59 (19.5)	19 (6.3)	4 (1.3)	0
Retired	17 (5.6)	2 (0.7)	3 (1)	1 (0.3)
Full or part time				
Full time	127 (64.1)	31 (15.7)	5 (2.5)	3 (1.5)
Part time	19 (9.6)	6 (3)	4 (2)	1 (0.5)
Student				
Yes	13 (4.3)	8 (2.7)	2 (.7)	0
No	210 (70.2)	48 (16.1)	14 (4.7)	4 (1.4)
Income				
$15,000 or below	27 (8.9)	9 (3)	3 (1)	0
$15,001–$30,000	60 (19.8)	15 (5)	4 (1.3)	3 (1)
$30,001–$45,000	55 (18.2)	16 (5.3)	2 (0.7)	2 (0.7)
Over $45,000	47 (15.5)	6 (2)	4 (1.3)	0
Refused	35 (11.6)	12 (4)	3 (1)	0
Sex				
Female	115 (37.3)	27 (8.8)	13 (4.2)	1 (0.3)
Male	112 (36.4)	32 (10.4)	4 (1.3)	4 (1.3)

Source: Courtesy of the NFLCP files.

TABLE 12.4 • DOMINANT CHARACTERISTICS AND IMPORTANCE OF ACCESS IN SUBSCRIBING TO CABLE

Dominant Characteristics of Subscribers	
Access not at all important	*Access of some importance*
Between the ages of 43 and 54	Between the ages of 43 and 54
Married	Single
College graduate	High school or less education
Employed full time	Employed full time
Income between $15,000–$30,000	Income between $30,000–$45,000
Female	Male

Source: Courtesy of the NFLCP files.

TABLE 12.5 • CROSS-REFERENCING OF SURVEY QUESTION 11 (WILLINGNESS TO PAY SAME) WITH SURVEY QUESTIONS 12–18 (DEMOGRAPHICS)

Marital Status		
Pay the same	*Single*	*Married*
Yes	28%	53%
No	6%	7%
Don't know	3%	3%

Education			
Pay the same	*High school*	*Some college*	*College*
Yes	27%	21%	33%
No	3%	5%	5%
Don't know	3%	1%	2%

Employed					
Pay the same	*Yes*	*No*	*Retired*	*Full time*	*Part time*
Yes	52%	22%	7%	67%	12%
No	11%	2%	0.7%	15%	2%
Don't know	3%	3%	0	3%	2%

(*continued*)

TABLE 12.5 • **CROSS-REFERENCING OF SURVEY QUESTION 11 (WILLINGNESS TO PAY SAME) WITH SURVEY QUESTIONS 12–18 (DEMOGRAPHICS) (continued)**

	Income			
Pay the same	15 or below	15–30	30–45	45 or over
Yes	12%	20%	21%	17%
No	1%	5%	2%	2%
Don't know	0	2%	1%	0

	Birthdate					
Pay the same	67–85	55–66	43–54	31–43	19–31	19 or below
Yes	9%	5%	15%	27%	15%	12%
No	3%	0	3%	4%	2%	2%
Don't know	2%	1%	0	1%	1%	6%

Source: Courtesy of the NFLCP files.

TABLE 12.6 • **CROSS-REFERENCING OF SURVEY QUESTION 10 (IMPORTANCE OF ACCESS IN SUBSCRIBING) WITH SURVEY QUESTION 11 (WILLINGNESS TO PAY THE SAME)**

	Willing to Pay the Same (Percent)		
Importance on subscriptions	Yes	No	Don't know
Not at all	64	6	4
Somewhat	13	5	1
Important	3	2	1
Very important	1	0.6	0

Source: Courtesy of the NFLCP files.

When the results from all of the cities were in hand, an attempt was made to relate them to the Kalamazoo survey. The researchers hoped to provide a picture of the national importance of public access and a composite description of its viewers. Although most of the surveys could not be directly correlated with the Kalamazoo study because the survey content did not directly correspond, certain access questions and responses were frequently alike.

The two most important questions posed in the Kalamazoo survey appear to be unique and therefore responses could not be correlated. These were the questions that asked whether access played a role in viewers' decisions to subscribe to cable, and whether those viewers would pay the same for basic cable services if access were no longer available. A uniform survey must be implemented nationally before any specific conclusions may be drawn.

From the surveys that were partially usable for correlation, one of the most frequently used questions was whether people were aware that their cable system had community access programming. Apparently most cable viewers were aware of access, because the response rate varied between 50% and 90% in the affirmative. Even more important is the percentage of viewers who have watched community access channels at one time or another. The response rate varied greatly, in a range of 30% to 78%, but a mean viewership of 59.9% was the average for ten cities.

Several surveys analyzed how often viewers watch access during an average week, and the most common response was a little more than once a week. This indicates that viewers do not watch access on a daily basis, but instead tune in for specific programs or for special events. Some of the surveys also asked what types of community access programs were watched. The most frequent responses were local sports events (usually associated with a high school or college), city council meetings, religious programming, and public affairs programming.

FIGURE 12.6 The Western Michigan University research team conducting the survey. Professor Frank R. Jamison in right foreground. Courtesy of the Western Michigan University Media Services and the NFLCP files.

In concluding the correlation, an attempt was made to determine the characteristics of an average access viewer. The results indicate that the viewer is female, between thirty and sixty in age, a high school graduate, married, employed, and with an income between $15,000 and $40,000. The sole difference between the national average access viewer and the average Kalamazoo viewer is that the latter tended to be a college graduate. The survey team agreed that a uniform survey needs to be done at the national level to get more information on the average access viewer and the habits of that viewer.

CONCLUSIONS

- Kalamazoo's survey of community access programming is statistically accurate, and its results present a factual picture.
- Community access is an important and widely used communication tool in the community.
- Community access plays a small but significant role in the subscription to cable television in Kalamazoo.
- A similar study should be initiated on a national basis, using this survey as a guideline.
- There is still much room for growth in community access, both locally and nationally.

ACCESS HIGHLIGHTS

We were going to subtitle this last chapter "Successes and Failures, Case Histories." Upon reflection, we have throughout the book pointed out what we thought were pitfalls in access. We felt from the outset that it was in the reader's best interest for us to candidly state that access has not been a rousing success throughout the United States. We sincerely believe that what you have read here will go a long way toward insuring that your access experience will be a positive one, and so we will conclude this book on a high note. And there are many, many access high notes. So many that it is very difficult to select which communities, which programs, or which situations to highlight in this chapter.

PRAISE FROM THE PRINT MEDIA

The print media have complimented access operations around the country. For example, we quote from an article in the *Kalamazoo Michigan Gazette* by radio and television writer Tom Haroldson. He was writing about election coverage by the broadcast stations in Kalamazoo and by the access center.

> *The cable access centers of Kalamazoo and Portage deserve a big hand for the way they handled Tuesday night's live coverage of elections in those cities. Viewers were treated to up-to-the-minute election results, timely, and sometimes insightful interviews, and all without having to wait until the 11 P.M. news. And by the time that the eleven o'clock news rolled around, it was already old news, and it didn't help that the local broadcast station's*

performance, delivering the results, was less than polished. The local broadcast station's interviews with mayoral winners from the two cities was a good idea, however, cable access viewers had already heard from those candidates and others. Most important, Kalamazoo and Portage voters were able to get cable access, and were willing to watch it, and didn't need to wait for whatever the local broadcast affiliate, or any other station wanted to deliver. It was rather enjoyable, flicking back and forth between the Kalamazoo and Portage access center coverages, catching this victory or that concession speech, the drama of upsets unfolding, the unhappiness of some losing candidates. I've said it before and it's worth repeating, Kalamazoo area TV viewers are lucky to have cable access coverage. Many cities, such as Detroit, don't have such luxuries, that can be used for constructive public service, while some people snicker, and call it amateur television. But Tuesday night's coverage was as slick and professional, particularly in Kalamazoo, as can be expected for any live, unrehearsed coverage.

Accolades like these about access news coverage from the local television press are the kinds of commentary that we see access operations receiving from honest journalists all across the country.

A HAWAII HIGHLIGHT

Another interesting story highlighting access comes from Honolulu, Hawaii. One might imagine that in Hawaii—a state best known for its tradewinds and gently swaying palm trees—the populace would be without problems, but of course this is not so. About three years ago, a group of concerned people in Honolulu formed an organization called Human Services Television (HSTV). As in many other states, community groups in Hawaii found that community television is a creative way to involve the general public in finding solutions to community problems. HSTV was created to offer health and social service providers a cost-effective vehicle for improving their educational and preventive outreach in the face of increasing budget cuts, inflation, and rising numbers of people requiring assistance.

The Office of Human Services of the city and county of Honolulu joined with the state's major planning and coordination agency for social services, the Health and Community Services Council of Hawaii, to develop a pilot video program depicting the plight of Honolulu's street people. The enthusiasm of the volunteers who worked on this initial video program and its usefulness to those concerned with the issue of homelessness inspired Human Services Television to move ahead with other projects. A steering committee was formed to coordinate the projects, and the Junior League of Honolulu encouraged HSTV to apply for a grant. The grant was awarded and it helped develop a video training program for volunteers and produce an entire year of programming. (See Figure 13.1.)

The programs produced by Human Services Television included "It's Up To Me," a program about teenage pregnancy in Hawaii. While producing this show, it became evident that the program would need to avoid adult judgements and adult presence if it was to be effective with adolescents. The program's goal was to develop a feeling that teenagers were speaking to their peers. The result was a half-hour show that presented perfectly candid adolescent conversations about abortion, adoption, and parenting experiences.

Another video program produced by HSTV that won numerous awards is called "The Honolulu Wheelchair Marathon." Producing the program presented a real logistical challenge. The program had to try to coordinate three video camera teams to follow four wheelchair racers over a 26.2-mile course. Moreover, the program intended to show the viewers the athletes as real people. The producers focused on the ability of the athletes and, implicitly, on the abilities of all of the disabled. HSTV won recognition for "The Honolulu Wheelchair Marathon" from the NFLCP, the National Cable Television Association, the Broadcast Media Confer-

FIGURE 13.1 Laotian refugee with HSTV crew member. Courtesy of the NFLCP photo files.

ence, the Hawaii International Film Festival, and the ITVA (International TV Association) of Honolulu. But most important, Human Services Television network has been a catalyst and advocate for community programming and has inspired others to use video to communicate their ideas and concerns.

A DALLAS HIGHLIGHT

In Dallas, Texas, local members of the American Association of Retired Persons (AARP) produced an access program called "Top of the Line." It all began in the spring of 1983, when the Dallas AARP staff and volunteer leaders in the area met to discuss recruiting members to learn video production. There was a lot of enthusiasm, but, initially, little participation. The craft was new to all of them, and most members of the community did not understand the concept of cable—let alone that of community access. But with the help of the access staff, a public meeting, and extensive publicity, a group of twelve older Americans was recruited and trained. Getting the first group going was the toughest barrier to overcome. Once this group learned the video production techniques and became certified access producers, the word spread.

Within a year, two crews of twenty-five members each had been trained, and they had produced a program that won an award in the National Federation of Local Cable Programmers Hometown USA Video Festival. The project now produces two video programs a month, with each crew responsible for producing a program. Gone are the long camera sweeps into space, the fumbling for the right button, the open mikes that should have been closed, and the voices coming from a dark set. The AARP access crew now laughs about the host who smiled more from fright than from friendliness. The crew also laughs about the time they neglected to push the record button on one of their first taped shows.

Dallas AARP's "Top of the Line" is usually in interview format, with a host and guests discussing topics of interest to older persons. The programs are built around a theme—for example, a health show might feature an arthritis specialist, a wellness expert, and a fitness demonstration, or it might concentrate on cardiovascular disease.

As of January, 1985, the AARP crews in Dallas have completed thirty-one shows targeting health, crime, housing, travel, consumer concerns, technology (i.e., older Americans and computers), retirement, music, nutrition, financial planning, telephone choices, creative employment, marriage after age sixty, and dancing. The success of this access project lies in the creative force of the people involved in its productions. When one crew member was asked what they get out of involvement with "Top of the Line," she replied, "ulcers and fear," and then added, "and a great deal of satisfaction." Another AARP producer said, "This is something completely new, something I never dreamed that I could do. That's what makes "Top of the Line" special."

A SAN FRANCISCO HIGHLIGHT

In San Francisco, California, a Boy Scout Explorer Post produces a regular program on the access channel called "Young Ideas," which began as a project to involve handicapped youngsters in television. It then evolved into a project for young people with social adjustment problems. Finally, in 1983, it was taken over by Explorer Post #400 in the San Francisco Bay Area Council of the Boy Scouts of America. The post is chartered to Viacom, the local cable company, and is assisted by The Public Eye, a local nonprofit organization dedicated to public access television. The Golden Gate Chapter of the American Red Cross lets the post use one of its rooms as a TV studio, and other groups, both private and public, in San Francisco donate "in kind" help such as blank videotapes, technical equipment, and editing facilities.

"Young Ideas" is produced entirely by youth. Post 400's Explorers fill all staffing positions, from the initial story conferences when program guests are chosen, through the technical slots, to on-camera talent. They operate the cameras, the audio gear, and set the lights; they direct, time, and edit the shows. They serve as host interviewers. Post adviser Keith Sinclair feels that "Young Ideas" gives these young people an opportunity to be responsible citizens without the possible stigma of failure. Each program they produce is an experiment in a highly disciplined structure, but in a fault-free atmosphere. This experience teaches life skills as much as career skills. Many of the scouts feel, however, that they want to go on from this experience to careers in television.

AN AUSTIN HIGHLIGHT

Access in Austin, Texas, has had a very long and hallowed career in access history. Austin Community Television (ACT), the access corporation, is one of the longest-operating nonprofit access corporations in the United States. As mentioned in earlier chapters of this book, nonprofit access management is the approach of choice of most community access groups.

Austin Community Television evolved in 1973, when a group of University of Texas students became interested in seeing television be something more than it has been to everyone in this country until recently. Most of all, they wanted a community channel: a channel on the cable system to be used for local programming.

ACT's first cablecast, in 1973, was unique. It involved driving a four-wheel drive vehicle to the top of a mountain outside of the city, where the cable company's head end was located. From that spot, placing their VCRs on the hood of the car, plugging power extension cords into receptacles in a shack at the mountain top, and playing back and cablecasting live, ACT started in the open air on the mountain top.

Today, Austin Community Television has grown into a fabulous operation that serves thousands of people in its home city. At least seventeen staff people are now employed by ACT, but it was a long journey to reach that number of employees. Up until four years ago, the access operation was very poorly funded. ACT relied on members of their Board of Directors to write grant proposals, and they worked with small grants that originated within the community.

In 1980–1981 Austin and the cable company began to renegotiate the franchise renewal. At that time, the cable company changed ownership, and at the same time Austin Community Television became a very large player in the franchise renewal process. (These negotiations took place during the period when the "gold rush" of franchising was in full blossom.) As a result of these situations during negotiations, Austin Community Television has received firm, reliable, and continuing funding.

All of the access prerequisites—staff, training, and outreach—have now been provided for. All of the diverse segments of the community, both the users and viewers, can participate in or watch ACT programming with the knowledge that vital funding is available for that programming.

Austin Community Television was honored at the 1985 NFLCP National Convention, where it was selected to receive the NFLCP Community Communications Award for access programming. This award is given to only one access center nationally each year. In addition, many programs produced by ACT won 1985 Hometown Video Festival awards.

One of the programs that has been cablecast on ACT's community access channels for over seven years is called "Alternative Views." This is a weekly public affairs program produced by Jim Morrow. This series has cablecast over 200 programs dealing with subjects and sources that are usually ignored or distorted in the traditional media. The program format is flexible to provide for the most effective presentation of the subject or subjects. Most shows include fifteen to twenty minutes of news gathered from a broad range of sources, including left-wing and right-wing publications, the business press, special-interest group newsletters and specialized journals. This material contrasts sharply with the treatment of the same subjects in the traditional media. The program may focus on one subject or several, and types of subjects presented vary.

Some of the programs contain interviews or presentations by well-known people. These personalities have included antinuclear activist Helen Caldicott, former US Attorney General Ramsey Clark, peace activist Daniel Ellsberg, civil rights lawyer William Kunstler, American Indian activist Russell Means, Nobel-prize-winning biologist George Wahl, child care specialist Dr. Benjamin Spock, and humorist and civil libertarian John Henry Falk. All of these people have been tremendously impressed by the treatment of their subjects and by the depth of knowledge that is imparted to viewers of "Alternative Views." John Henry Falk said, "A great number of citizens in the Austin area look to this show as the most exciting thing that happens on television. "Alternative Views" has my overwhelming enthusiasm. It's what television really should be all about." And Professor Carl Jensen of Sonoma State University told ACT, "You are providing Austin Community Television viewers with some of the best documentary news programming in America." "Alternative Views" is a major success story in public access programming. It is now being seen on numerous access channels all across the United States, and it is providing a *real* alternative news program.

AN ATHENS, GEORGIA, LEASED ACCESS SUCCESS STORY ___

In Athens, Georgia, early in 1981, an experiment was started using a leased access channel. About ten years earlier, Chuck Searcy and several other young entrepreneurs had taken over the *Athens Observer,* a newspaper in that city. During that ten-year period, the *Observer,* carved out a niche in a very competitive media market. But, surprisingly in a city of 70,000, there was no local commercial television station. The university station, which was part of the state educational network, offered little or no local programming.

Searcy and his partners at the *Athens Observer* saw an opportunity to use the newspaper's good reputation and its advertising staff to reach into a new market. That new market was local television programming, and they called their idea Observer Television.

Their initial meeting with the cable company's local system manager was not particularly positive. They described to him their plans for local television and suggested negotiations for the use of a local cable channel. The cable company manager's response was an immediate and direct "no." He pointed out that each channel on the system represented a significant financial investment and that his company was not prepared to turn over a channel to anyone unless there was a clear benefit to the cable company. It was strictly business.

About a week later, Searcy and an associate again met with the cable operator's local manager, as well as with several regional and corporate officers of the cable company. The newspaper men explained their concept of Observer Television, which was based on the need for local information to complement the existing array of network and satellite program services available through cable. In exchange for the use of one channel on the cable system, the cable operator would reap several benefits. These benefits included increased good will in the community toward the provider of local television, an outlet for public access requests, an outlet for promotional campaigns to sell additional cable services, and revenues from the channel lease. All of these benefits would come at little or no cost to the cable company.

Searcy's company, Observer Television, offered to buy all of the needed equipment, provide a staff and studio facilities, ensure local programming, handle advertising sales, and be completely accountable and responsible for all aspects of the operation of the leased channel. The offer left the cable company free to do what it did best: maintain and improve the technical capabilities of the system and provide more and better services for its customers—which, in turn, would expand its customer base and increase its revenues and profits.

The contrast between the first and second meetings of the two groups was striking. This time the cable company officials listened carefully and seemed to like the proposal.

Observer Television began operations in April, 1983, using only a character generator. Throughout that summer, cameras, VCRs, lights, and other equipment arrived as a former bookstore was renovated to become the studio and production facility. Observer Television carried three local newscasts: one in the morning, a live one-hour "Lunchbreak," and another hour of live programming called "Around Athens" from 7:00 to 8:00 PM every weekday.

It was not until the fall of 1983 that Observer Television began to make its mark as a local station. By that time most of its equipment was in place, and the station had a crew of program and production people (including three full-time people), half a dozen part-time volunteers, and interns from the University of Georgia. That fall, Observer entered into an agreement with the local school system to carry television replays of all home and away high school football games. These games vividly demonstrated the impact of local cable television to the local cable audience.

To do the football shoots, the production manager loaded virtually all of the studio and control room equipment into the back of a van. The games announcer was a local radio professional who soon became the sales manager for Observer Television. Sponsorship of the football games was sold out in a matter of days, and the major sponsors for that season included Coca Cola, Toyota, First National Bank, and Arby's of Athens.

Observer Television is a prime example of one of the less common forms of access—leased access. Over the last several years it has continued to expand its programming and to provide a very special kind of television that is not received in any other way by the people of Athens.

A SAN DIEGO HIGHLIGHT

Since December 1981, the San Diego, California, County Department of Social Services has been producing a public access program called "Social Services Gamut." This weekly self-help television series is cablecast on four access channels in the San Diego area and reaches 400,000 homes. "Social Services Gamut" focuses on topics such as child abuse, adoption, nutrition, and money management. The show embraces a wide range of personal and social problems. It was not conceived as a program that viewers would choose to watch weekly; instead, its creators anticipated a constantly changing, very sporadic audience of individual viewers who selected only the shows that directly related to their lives. (See Figure 13.2.)

However, a recent survey about "Social Services Gamut" unexpectedly found that there was a core of regular viewers. Two independent market research firms measured the audience on the four cable systems that carry

FIGURE 13.2 Trainer pins a microphone on the moderator of "Social Services Gamut." Courtesy of the NFLCP files.

the program. The results showed that "Gamut" was identified by 41% of the viewing audience on one cable system and 30% of the viewing audience on another system. One third of the audience watched the program at least twice a month, an additional 40% tuned in once or twice a month, and 26% watched "Social Services Gamut" once a month or less. This program has almost limitless use in providing training and increasing education. In an era of severe monetary constraints on social services, "Gamut" is an extremely effective method of reaching the community with continuing information on the available social services.

CONCLUSION

In this chapter we have highlighted just a very few of the success stories from the world of access programming. There are many more that we know about, and many, many more of which we are unaware. Access is a new kind of television, one that can provide public service information, political information, and, in general terms, a window on the community. We invite you to open that window.

APPENDIX A

SURVEY

OF

CABLE TV UTILIZATION

by

COLLEGES and UNIVERSITIES

Instructional Television Services
The University of Texas at Arlington
P.O. Box 19378
Arlington, TX 76019

Robert G. McCartney
Director, ITVS
Ha Phan
Research

July 1985

Survey of Cable Utilization
Colleges and Universities

Introduction

During the Spring of 1985, Instructional Television Services, University of Texas at Arlington undertook to survey all U.S. institutions of Higher Education which were using cable television. The impetus for the survey came from the many requests to UTA for information on how our own cable casting operation was organized, funded and specifically how students were integrated into the over all operation.

Many of the requests from other institutions were referred to us by the National Federation of Local Cable Programmers (NFLCP), an organization of public access cable users, cable operators, educational and governmental users and special interest groups which use local cable TV. Some, but not all, of the requestors belong to NFLCP. As our involvement with cable grew, so also did our participation in NFLCP, both nationally and regionally. In parallel, our contacts with other educational institutions interested in cable grew. From these contacts it became obvious that:

1. Cable operations at colleges and universities were being developed locally with a variety of organizational structures.
2. Each institution was making ad hoc isolated decisions on policies and procedures.
3. There was no national forum, outside of NFLCP, in which collegiate cable leaders could meet and exchange experience and ideas.

CONTENTS

Hypotheses

Our cursory subjective knowledge of the use of cable by colleges and universities, gained in over six years in this work, indicated certain basic assumptions we could make about this area of endeavor. We felt that the survey would lend statistical credence to our initial hypotheses. Our assumptions to be tested were:

1. Cable operations were minimally budgeted.

2. Successful operations were notable for a high degree of student involvement.

3. Cable operations were integrated with academic work in student participation.

4. Cable operations did not have a large number of fulltime staff.

5. Most institutions had not been in cable for a long period of time.

6. Cable programming, nation-wide, would be about equally divided between instructional (credit) courses, and general programming for student audiences.

7. Cable casting did not actually represent a large number of hours a week. (Arbitrarily, less than 10 hours.)

8. Cable operators are minimally equipped, and usually share facilities.

9. Most cable operators are under the jurisdiction of an academic department.

The actual survey results, proved the veracity of most of our hypotheses. Exceptions showed up in statements 1., 5., and 9. of the hypotheses as outlined in the analysis following.

(One of our contacts used the analogy of each of us, "reinventing the wheel," as institutions individually struggled to come-on-line with some practical utilization of cable TV.)

A search of the literature found no specific studies of our particular application of cable. (The NFLCP provides some excellent, if dated, case studies of utilization by public schools.)

We therefore thought that a comprehensive survey would be of practical value to address the situation found in points 1-3, above. It would provide some sort of statistical base upon which to assess the local cable operation and stimulate interaction between those cable specialists who are practicing in comparative isolation from their counterparts in other institutions.

Method

The survey instrument (questionnaire, attached) was designed to be objective, easy to complete, and easy to analyze statistically. We included what we considered to be minimal significant information with a view to eliciting as large a response as possible. Since the survey was to be conducted with available personnel and resources, not underwritten in anyway, we kept to a direct incisive design, within our capabilities. A small concession to respondents who wished to include more detail was included as question 27.

Development of the mailing list turned out to be a major effort as there was no complete directory listing available for our specific group. We used the Knowledge Industries publication, The Video Register, Broadcasting Yearbook, and our own list of correspondents to identify our target group. We also received assistance from Association of Higher Education, which became interested in our survey. (Ms. Joan Gudgel, AHE, Center for Learning and Telecommunications, provided printed mailing labels for some 100 institutions.) Other Campus Network affiliates on cable rounded out our compilation of the mailing list. Some 400 identifiable institutions were included.

An initial mailing, with cover letter (attached), was sent in mid-March with a follow up to nonrespondents in April. The survey netted a noteable 44% response.

A complete tabulation of responses is attached, together with the specific questions on the survey form.

4

Survey Results

1) The response rate for this survey is 43.9% or 173 out of 394 institutions surveyed. 63 out of 173 do not cablecast resulting in 111 institutions applicable to our survey, yielding a net response rate of 28.17%. The following analysis is therefore based on 111 surveys.

2) More than half of the institutions surveyed have been cablecasting within a period of 0 to 4 years. A noticeable number of institutions have operated over 8 years (21.63%).

3) More than half constitute an integral part of a service (administrative) unit and about a third is part of an academic department. Others encompass the above areas or are part of continuing education or instructional media centers.

4) Titles for responsible supervisors are various. They are:

 a) Director (of TV, Telecommunications)
 b) President/Vice President/Executive V.P.
 c) Chairman
 d) Dean (of Academic Affairs)/Associate Dean
 e) Manager
 f) Department Head
 g) Provost
 h) Vice Chancellor
 i) Media Advisor

"Director" title is most frequently used; the rest is evenly distributed.

5) 56.76% have a dedicated access channel specified in the franchise.

6) Half have published a program schedule.

5

7) Dominant cablecasting period is from 6 p.m. to 12 p.m. and from noon to 6 p.m. About 1 out of 3 cablecast on Saturday and Sunday.

8) During weeks when classes are in session, 36% cablecast less than 10 hours a week and 28% cablecast more than 50 hours.

9) During weeks when classes are not in session, most institutions cablecast less than 10 hours per week. About 1 out of 4 cablecast between 11 to 40 hours per week.

10) Most respondents have an annual budget below $50,000. Approximately 14% have an annual budget over $100,000. (Our survey refers to "overall" annual budget; respondents debated themselves, with capital or operating budgets" resulting in a high percentage of missing answers (17%)).

11) Studio facilities are predominantly shared with academic activity (62.16%). A small percentage is provided by cable operators or separate entities.

12) About one-half have less than 5 cameras, and one-half have between 5 and 12. They are almost entirely of Broadcast Quality or High-End Industrial. Nearly two-thirds have 10 to 19 VTR's. Approximately 15 institutions have more than 19 VTR's, up to a maximum of 40. They are predominantly 3/4" Umatic and 1/2" VHS.

13) Most respondents allocate more than a quarter of the budget to personnel salary which includes professionals, classified salaries and student wages.

14) Students play a dominant role in production. In more than half of the institutions surveyed, the percentage of student participation is between 76% to 100%. However, the percentage of supervisory or administrative responsibilities assumed by students lies between 0 and 25%.

15) More than half of the respondents produce work under contract with external agencies and only very few (12.61%) sell commercial time on the access channel. (We asked this question to ascertain whether the activity had made any effort to generate operational funds from their own resources.)

16) News, sports and talk shows are the most usual locally produced programs. Telecourses rank next. Other programs include instructional movies, live call-ins, How-To-Demos, documentaries, lectures, seminars, special events, health, arts, student video essays and documentaries.

17) 45% of the respondents have no newscasts. 1 out of 2 have one to five per week and only a few have six or more (3%).

18) Of the total cablecast program hours, 38% of the respondents have 1 - 25% locally produced programs

A-1

The University of Texas at Arlington

Instructional Television Services (UTA)
Channel 24
Box 19378
Arlington TX 76019
(817) 273-2905

March 18, 1985

Dear Colleague:

Many of us in Higher Education are using access television for a variety of purposes. The attached survey represents our effort to develop a coherent picture, nationally, of the broad spectrum of our various applications of cable by colleges and universities. The occasion of the survey is also an initial step towards putting us in closer contact with each other.

We would greatly appreciate your taking the few minutes required to complete the survey and return it by April 12, 1985 in the self-addressed, postage paid envelope provided.

UTA has a dedicated access channel which is specified in the franchise with Arlington Telecable. We use it to present general programming to a student/community audience. We are currently producing 8 hours of programming Monday through Friday. Students run the cablecasting station, with over 100 participating each semester. Though a student activity, the cable operation is closely integrated with the curriculum of the Communications Department. (The attached brochure gives an outline of the structure and functions of ITVS, which guides the cable operation.)

We are members of the National Federation of Local Cable Programmers (NFLCP). This organization was instrumental in developing the mailing list for this survey.) We've met many of you at the NFLCP National and Regional Conferences.

However, for our initial research on the survey, we're aware that there are many cable practitioners from Higher Education, that are not NFLCP members. Though our work in cable is a highly specialized application of communications technology, we should not operate in a vacuum, unaware of the many creative-innovative things that are being done by our counterparts across the country. We, at ITVS, have always felt that the NFLCP is the logical forum, which could meet exchange mutual support and ideas. We will publish the results through NFLCP and further-more, we are planning on presenting the results of our survey at the NFLCP Convention in Boston this coming July. It would be nice to meet you there.

Let us thank you in advance for your time and attention in completing and returning the survey.

Yours truly,

Robert G. McCartney
Director

cc: Sue Miller Buske
Executive Director NFLCP
906 Pennsylvania Ave. SE
Washington, D.C. 20003

8

and nearly 28% have 100% locally produced programs.

These latter tend to be the operations that have the fewest program hours per week.

19) The most frequent source for outside program material was other university productions (41.44%). Other important sources include commercial productions (i.e., Campus Network, Rockworld, music industry productions, licensed telecourses, satellite services, and state public television services).

20) 56.76% of the respondents have an average of 0 to 2 repeats for each individual program each week. And 31% have an average of 3 to 4 per week. Only a small percentage have more than 4 replays per week.

21) Students in residence on campus and community are both primary target audiences.

22) 83.78% of the respondents do not have figures on the penetration of student viewers.

23) Most institutions define the uniqueness of their operation due to:

a) level of student participation: good students, low budget, limited staff, large number of volunteers. Financial support from the student body. (Example: Per head student fee of $1.00)

b) Program diversification: repeat of cable cast lessons has witnessed a sharp increase (40%) in student enrollment (live programs produced by students).

c) Equipment at the facility: home computer based automation system, separate studios and control rooms—production of student projects and main station can be simultaneous. High quality production faculty and field production equipment with broadcast engineer available, remote van, edit systems provided by cable franchise, convergence editors (one response).

A-2

TABULATION OF RESPONSES

INSTRUCTIONS: Please circle one response for each item unless
otherwise indicated.

	NUMBER	PERCENTAGE

1. How long ago did your institution start cablecasting?

a. Less than 2 years	a) 22	19.82%
b. 2 - 4 years	b) 42	37.83%
c. 5 - 6 years	c) 13	11.71%
d. 7 - 8 years	d) 10	9.01%
e. more than 8 years	e) 24	21.63%

2. Your organization is an integral part of:

a. an academic department	a) 33	29.74%
b. a service (administrative) unit	b) 63	56.76%
c. other (Please specify): _____	c) 15	13.5%

3. To whom does your cable access supervisor report?

_____
Title

4. Do you have a dedicated access channel for your institution
which is specified in the franchise?

a. Yes	a) 63	56.76%
b. No	b) 48	43.24%

5. Do you publish or have published a program schedule?

a. Yes. If so, please enclose publish	a) 57	51.35%
b. No	b) 53	47.75%

6. Cablecasting occurs:

	Monday	Tuesday	Wednesday	Thursday	Friday	Sat.	Sunday		
6am-noon	*	*	*	*	a) 50 *	*	*	(d)	36
noon-6pm	*	*	*	*	b) 70 *	*	*	(e)	40
6pm-12pm	*	*	*	*	c) 89 *	*	*	(f)	38

Number of responses for
each time period.

7. During weeks when classes <u>are</u> in session, how many hours
per week do you cablecast?

a. Less than 10	a) 40	36.03%
b. 11 - 20	b) 17	15.31%
c. 21 - 40	c) 18	16.22%
d. 41 - 50	d) 4	3.61%
e. more than 50	e) 32	28.83%

8. During weeks when classes are <u>not</u> in session, how many hours
per week do you cablecast?

a. Less than 10	a) 55	49.55%
b. 11 - 20	b) 13	11.71%
c. 21 - 40	c) 14	12.62%
d. 41 - 50	d) 4	3.60%
e. more than 50	e) 18	16.21%
MISSING	7	6.30%

A-3

		NUMBER	PERCENTAGE

9. What is your annual budget for the operation?

		NUMBER	PERCENTAGE
a. Less than $10,000	a)	33	29.73%
b. $10,001 - $20,000	b)	14	12.61%
c. $20,001 - $50,000	c)	18	16.22%
d. $50,001 - $100,000	d)	13	11.71%
e. Over $100,000	e)	15	13.51%
	MISSING	18	16.22%

10. Are your studio facilities:

a. exclusively for your cable operation	a)	6	5.40%
b. shared with academic activity	b)	69	62.16%
c. provided by cable operator (i.e. Local Origination)	c)	7	6.31%
d. other (Please specify): _____	d)	22	19.82%
_____	MISSING	7	6.31%

11. Please tell us about your equipment.

Cameras: Please provide total number available _____
They are: (Check all that apply)

a. Broadcast Quality	a)	41
b. High-end industrial	b)	70
c. Low-end industrial	c)	24
d. Consumer grade	d)	15

VTR's: Please provide total number available_____
They are: (Check all that apply)

a. 2" Quad	a)	12
b. 1" Format B or C	b)	15
c. 3/4" Umatic	c)	95
d. 1/2" VHS	d)	75
e. 1/2" BETA	e)	31
f. Other (Please specify): _____	f)	3

12. How many full-time professional staff members ⊔ employ?

a. Less than 5	a)	74	66.67%
b. 6 - 10	b)	18	16.23%
c. 11 - 15	c)	1	0.9 %
d. 16 - 20	d)	2	1.8 %
e. more than 20	e)	4	3.6 %
	MISSING	12	10.8 %

13. What portion of the budget is allocated to personnel? (including professionals, classified salaries and student wages)

a. 0 - 15%	a)	23	20.73%
b. 16% - 25%	b)	7	6.31%
c. 26% - 50%	c)	25	22.52%
d. more than 50%	d)	38	34.23%
	MISSING	18	16.21%

14. What percentage of production is done with student participation?

a. 0 - 25%	a)	27	24.32%
b. 26% - 50%	b)	6	5.40%
c. 51% - 75%	c)	14	12.62%
d. 76% - 100%	d)	53	57.66%
	MISSING	11	

15. What percentage of supervisory or administrative responsibilities is assumed by students?

a. 0 - 25%	a)	74	66.67%
b. 26% - 50%	b)	8	7.20%
c. 51% - 75%	c)	12	10.81%
d. 76% - 100%	d)	6	5.40%
	MISSING	11	9.92%

A-4

		NUMBER	PERCENTAGE

16. Do you produce work under contract with external agencies?

    a. Yes   a) 59   53.15%
    b. No   b) 52   46.85%

17. Do you sell commercial time on your access channel?

    a. Yes   a) 14   12.61%
    b. No   b) 97   87.39%

18. Locally produced programs include (Please check all that apply)

    a. Telecourses   a) 49   44.14%
    b. News   b) 62   55.86%
    c. Sports   c) 62   55.86%
    d. Music Video Programs   d) 28   25.22%
    e. Talk Shows   e) 77   69.37%
    f. Other (Please specify) _____   f) 41   36.94%
    _____

19. What is the number of newscasts per week?

    a. none   a) 50   45.04%
    b. 1 - 2   b) 36   32.43%
    c. 3 - 5   c) 21   18.92%
    d. 6 or more   d) 4   3.61%

20. Of your total cablecast program hours, what percentage is
locally produced?

    a. none   a) 6   5.40%
    b. 1 - 25%   b) 42   37.84%
    c. 26% - 50%   c) 13   11.71%
    d. more than 50%   d) 19   17.12%
    e. 100%   e) 31   27.93%

21. What types of outside program materials do you utilize?
(Check all that apply)

    a. Music industry production   a) 14   12.61%
    b. Other-university-productions   b) 46   41.44%
    c. Religious programs   c) 8   7.20%
    d. Commercial productions (Source: _____)   d) 27   24.32%
    e. Other _____   e) 34   30.63%
    f. None of the above   f) 23   20.72%

22. What is the average number of repeats (replays) of each
individual program each week?

    a. 0 - 2   a) 63   56.76%
    b. 3 - 4   b) 35   31.53%
    c. 5 - 6   c) 7   6.30%
    d. more than 6   d) 4   3.60%
                                    MISSING   2   1.81%

23. Your target audience is:

    a. students in residence on campus   a) 10   9.01%
    b. community   b) 49   44.14%
    c. both   c) 52   46.85%

A-5

24. Do you have figures on the penetration of student viewers?

    a. Yes
    b. No

25. What define(s) the uniqueness of your operation?

    a. Program diversification
    b. Budget size and allocation
    c. Equipment at the facility
    d. Level of student participation
    e. Technical manpower support
    f. Other. Please specify. _____
    _____
    _____

	NUMBER	PERCENTAGE
a)	18	16.22%
b)	93	83.78%
a)	35	31.53%
b)	21	18.92%
c)	31	27.92%
d)	42	37.83%
e)	17	15.31%
f)	24	21.62%

26. Please briefly discuss the areas of uniqueness that were checked in question 26.

    _____
    _____
    _____
    _____
    _____
    _____
    _____

27. Please enclose any printed materials concerning your operation.

    THANK YOU FOR YOUR COOPERATION IN COMPLETING THIS SURVEY

    PLEASE RETURN YOUR COMPLETED QUESTIONNAIRE TO INSTRUCTIONAL TELEVISION SERVICES, UTA BOX 19378, ARLINGTON, TX 76019 BY <u>APRIL 12,1985</u>.

APPENDIX B
ACCESS RESOURCES

TRADE AND SERVICE ORGANIZATIONS

Association of Independent Video and Filmmakers, Inc.
625 Broadway
New York, NY   10012
(212) 473–3400

Board of Cooperative Educational Services
New York State Educational Department
Bureau of Educational Communications
Room 325, Educational Building
Albany, NY   12234

Cabletelevision Advertising Bureau
767 Third Avenue
New York, NY   10017
(212) 751–7770

The Cable Television Information Center
1800 North Kent Street—Suite 1007
Arlington, VA   22209
(703) 528–6846

Independent Cinema Artists and Producers
625 Broadway
New York, NY   10012
(212) 522–9183

The International City Management Association Cable Committee
1120 G Street NW
Washington, DC   20005
(202) 626–4600

National Assembly of Media Arts Centers
c/o New York Foundation for the Arts
5 Beekman Street, Room 600
New York, NY   10038
(212) 233–3900

The National Association of Telecommunications Officers and Advisors
National League of Cities
1301 Pennsylvania Avenue NW
Washington, DC   20004
(202) 626–3020

National Cable Television Association
1724 Massachusetts Avenue NW
Washington, DC   20036
(202) 775–3550

National Federation of Local Cable Programmers
906 Pennsylvania Avenue SE
Washington, DC   22203
(202) 544–7272

Office of Communication
United Church of Christ
105 Madison Avenue
New York, NY   10016
(212) 683–5656

Women in Cable
2033 M Street NW—Suite 703
Washington, DC   20036
(202) 296–7245

US GOVERNMENT

Federal Communications Commission
Complaints and Information Branch
Cable Television Bureau
Washington, DC   20554

PUBLICATIONS

Cablevision
Titsch Publishing Company
1130 Delaware Plaza
PO Box 4305
Denver, CO   80204

Channels of Communication
1515 Broadway
New York, NY   10036

Community Television Review
National Federation of Local Cable Programmers
906 Pennsylvania Avenue SE
Washington, DC   20002

CTIC CableReports
Cable Television Information Center
1800 North Kent Street
Arlington, VA   22209

Multichannel News
Fairchild Publications
1762–66 Emerson Street
Denver, CO   80218

Television Factbook
Television Digest, Inc.
1836 Jefferson Place NW
Washington, DC   20036

CONSULTANTS

This list of consultants is not meant to be complete. It presents a cross-section of consultants working in the areas indicated.

Franchising, Renewal, Change of Ownership

CTIC Associates
Harold Horn, President
1800 North Kent Street
Arlington, VA   22209

Malarkey Taylor & Associates
1301 Pennsylvania Avenue NW #200
Washington, DC   20004

Miller & Young
1150 Connecticut Avenue NW #617
Washington, DC   20036

Preston Thorgrimson Ellis & Holman
1735 New York Avenue NW #500
Washington, DC   20006

Rice Associates
1346 Connecticut Avenue NW #325
Washington, DC   20036

Spiegel & McDiarmid
1350 New York Avenue NW #1100
Washington, DC   20005

Local Cable Programming Management, Policy, and Implementation

National Federation of Local Cable Programmers
906 Pennsylvania Avenue SE
Washington, DC   20003

Rice Associates
1346 Connecticut Avenue NW #325
Washington, DC   20036

INDEX